Low-Power VLSI Circuits and Systems

Low-Power VLSI Circuits and Systems

Editor

Hasad Demirci

scitus
academics

Low-Power VLSI Circuits and Systems
Edited by **Hasad Demirci**

ISBN: 978-1-68117-160-9
Library of Congress Control Number: 2015953032

SCITUS Academics LLC,
www.scitusacademics.com
Box No. 4766, 616 Corporate Way,
Suite 2, Valley Cottage,
NY 10989

Notice

Printed in United States of America on Acid Free Paper

Preface

Very-large-scale integration (VLSI) is the practice of creating an integrated circuit by combining thousands of transistors into a single chip. VLSI began in the 1970s when complex semiconductor and communication technologies were being developed. The microprocessor is a VLSI device. Before the introduction of VLSI technology most integrated circuits had a limited set of functions they could perform. An electronic circuit might be made of a CPU, ROM, RAM and other glue logic. VLSI lets integrated circuit designers add all of these into one chip.

Low-power VLSI circuits and systems delivers a wide-ranging coverage of different aspects of low power circuit synthesis at numerous levels of design hierarchy; starting from the layout level to the system level. It addresses both process technologies and device modeling. Power dissipation in CMOS circuits, several practical circuit examples, and low-power techniques are discussed. The book also delivers an extensive study of advanced CMOS subsystem design. This book is a reference tool useful for students, researchers, as well as practicing engineers.

Table of Contents

CHAPTER 1

Design of Ultra-Low Power PMOS and NMOS for NANO Scale VLSI Circuits

Ashok Babu Ch[1], J. V. R. Ravindra[2], K. Lalkishore[3]

[1]Department of Electronics and Communication Engineering, SVIT, Sec-Bad, India
[2]Department of Electronics and Communication Engineering, Vardhaman College of Engineering, Hyderabad, India
[3]Jawaharalal Nehru Technological University, Anantapur, India

ABSTRACT

CMOS devices play a major role in most of the digital design, since CMOS devices have larger density and consume less power. The integrated circuit performance mostly depends on the basic devices and its scaling methods, but in conventional CMOS devices in ultra-deep submicron technology, leakage power becomes the major portion apart of dynamic power. The demerits of the conventional CMOS is less speed and, more leakage, for any digital design PDP is the figure of merit which can be used to determine energy consumed per switching event, hence we designed a NOVEL NMOS and PMOS which has superior performance than conventional PMOS and NMOS, the design and performance checked at 90 nm, 180 nm and 45 nm technology and calculate the performance values.

INTRODUCTION

The need for low power chips is the increased marked demand for portable consumer electronics powered by the batteries, which have not experienced the similar rapid density growth compared to electronics circuits. It has been concluded that battery technology alone will not solve power problems in the near future. Therefore designing a chip with low power is becoming a major concern in the chip design industry. The different power reduction techniques from device level to circuit level and system level have been shown as a low power taxonomy [1].

In ultra-deep submicron technology especially below 45 nm technology, leakage power is becoming a major concern than the dynamic power. As the technology is going on scaling down, and we are now in 20 nm/15 nm/ 10 nm technology proper controlling of leakage power is becoming challenge factor.

Any power reduction techniques at device level will certainly lower the overall power consumption at system level (power reduction means it should be applicable from device level to system level, there is no guarantee that device level power reduction always applicable at system level) [2].

Related work: The existing numerous power reduction techniques in the field of VLSI are suitable for only 90 nm and above, but continuous scaling of the device in the present VLSI technology many of them may not be suitable. Moreover if the technique is suitable for power reduction than it fails for high speed, hence a new technique is required that is suitable for both power and speed [3].

Sources of power dissipations are

$$P_{\text{avg}} = P_{\text{static}} + P_{\text{dynamic}} + P_{\text{leakage}} + P_{\text{shortckt}}.$$

The major source in the overall power dissipation is the dynamic power

$$P_{\text{dynamic}} = \alpha \times V_{DD}^2 \times C_L \times f.$$

α = transition activity factor, V_{DD} = supply voltage, C_L = load capacitance, f = frequency of operation.

As the technology shrinks leakage power is to be dealt very carefully.

TAXONOMY OF LOW POWER

In Figure 1 different power dissipation and power reduction techniques are clearly shown, all these techniques are applicable in low power design [4] .

PROBLEM STATEMENT

Here an attempt has been made to control leakage power by proposed PMOS and NMOS, the fundamental limit of conventional PMOS and

NMOS is higher leakage and lower sped in deep submicron technology sine power

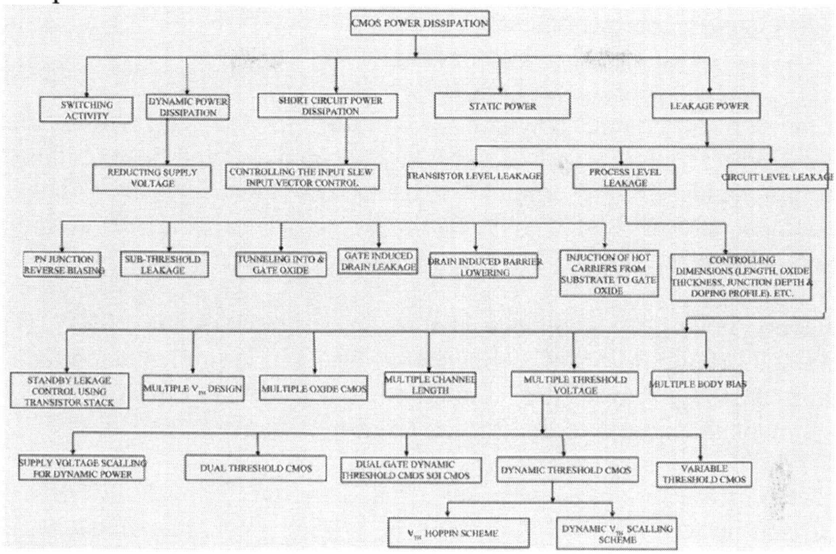

Figure 1. Taxonomy of low power.

supply reduction lowers three times the threshold voltage will degrade the speed of the circuit considerably, as we are striving forward to higher level of integration, area is not having that much of concern, now we are in position to keep millions of transistors in a single die, the other performance parameters like power, speed and delay moreover power delay product are important parameters which can decide the role of the device in nano scale regime [5]. In design abstraction we are approaching bottom up design as shown in the below Figure 2, basically CMOS device level analysis is done and comparative study is done among different MOS devices.

Demerits of Conventional PMOS & NMOS

Figure 3(a) and Figure 3(b) show the conventional PMOS and NMOS and having the following demerits.

- Lower speed.
- Higher leakage in deep submicron technology.

Figure 4 in DT PMOS and DT NMOS body is connected to gate terminal of the MOS device and as the gate voltage is changing body biasing takes place.

Working of proposed PMOS and NMOS: in Figure 5 for top transistor (PM0) a reference voltage of 0.8 V is applied, and input is applied to the gate terminals of bottom transistor (PM2) and output is taken across the drain terminals of the right transistor (PM1) [6].

Both the conventional PMOS and NMOS, DT PMOS, DT NMOS and proposed PMOS and proposed NMOS are implemented using 45 nm technology and all are simulated using Cadence Virtuoso Design Environment and simulated and results are shown Table 1. From Table 1 the proposed PMOS and NMOS have low PDP value.

Comparison table for the conventional PMOS and NMOS, DT PMOS, DT NMOS and proposed PMOS and proposed NMOS are implemented using 90 nm technology and all are simulated using Cadence Virtuoso Design Environment and simulated and results are shown Table 2.

Comparison table for the conventional PMOS and NMOS, DT PMOS, DT NMOS and proposed PMOS and proposed NMOS are implemented using 180 nm technology and all are simulated using Cadence Virtuoso Design Environment and simulated and results are shown Table 3.

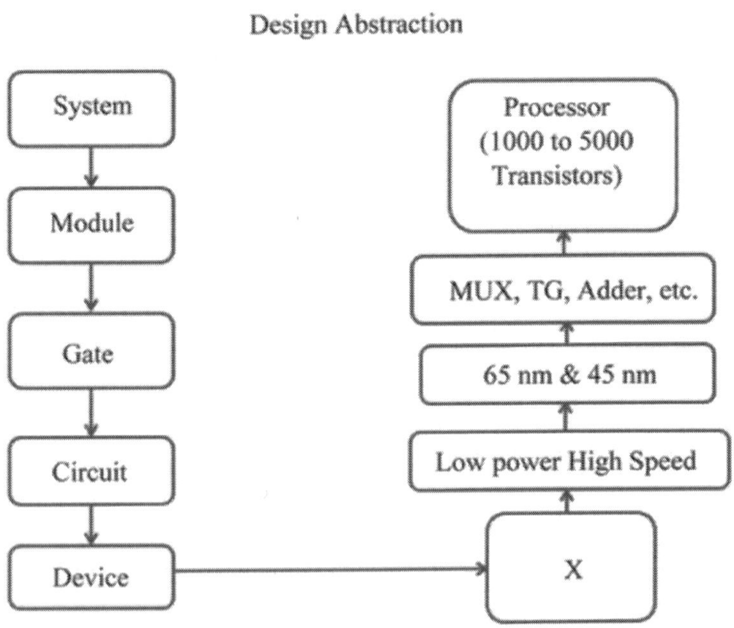

Figure 2. Design abstraction.

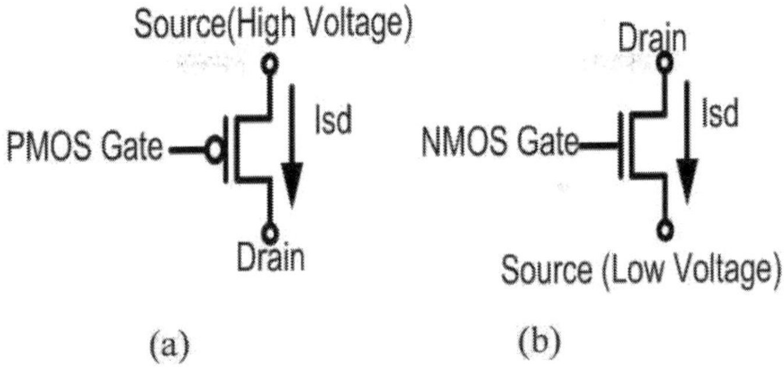

Figure 3. Conventional PMOS (a) & NMOS (b).

Figure 4. DT PMOS and DT NMOS.

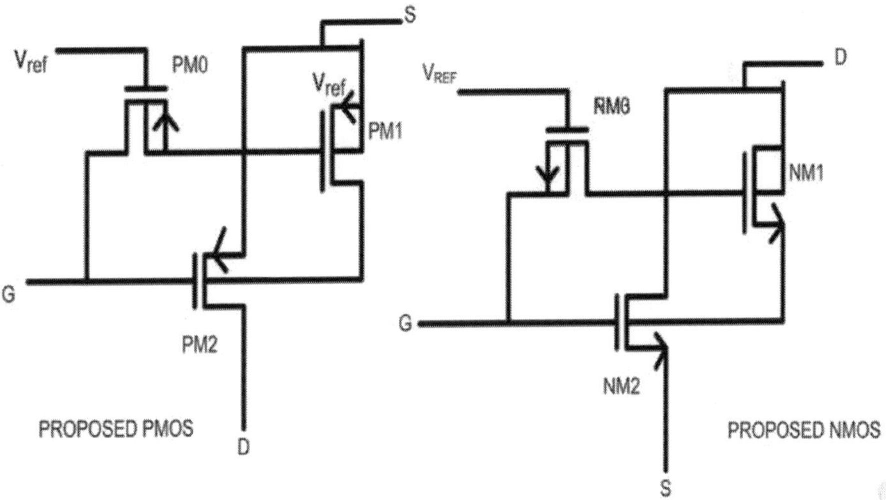

Figure 5. Proposed PMOS and NMOS.

Table 1. 45 technology

Device	Output max value	Out min value	Avg. power	Delay	Static power		PDP
					$V_{IN} = 0$ V	$V_{IN} = 1.1$ V	
PMOS	1.1 V	0.323 V	5.37×10^{-9}	16 ps	116×19^{-15}	481×10^{-24}	8.59×10^{-20}
DTMOS	0.805 V	0.217 V	5.85×10^{-3}	4 ns	588×10^{-63}	11.26×10^{-3}	2.34×10^{-14}
NOVEL PMOS	1.1 V	0.250 V	1.71×10^{-9} V	4 ps	1.443×10^{-12}	822.5×10^{-15}	6.84×10^{-21}
NMOS	0.79	0	4.412×10^{-9}	13 ps	0	101.1×10^{-15}	5.73×10^{-20}
DTMOS	0.885 V	0.277 V	2.906×10^{-3}	4 ns	6.24×10^{-3}	48×10^{-27}	2.6154×10^{-14}
PROPOSED NMOS	0.818 V	0 V	6.29×10^{-9}	9 ps	151.8×10^{-15}	268.8×10^{-15}	5.661×10^{-20}

The PMOS and NMOS are good transfer of 1 and good transfer of 0, the comparative study is done on various types of MOS devices like PMOS, DTMOS and proposed MOS devices, the average power, delay and PDP is calculated by using the cadence virtuoso design environment tool for 45 nm, 90 nm, 180 nm technology [7].

CONVENTIONAL CMOS INVERTER

Figure 6 shows a conventional CMOS inverter which is simulated using Cadence Virtuoso Design tool and simulated outputs are shown in Figure 7, Figure 8 shows CMOS output when all Zero's Applied and Figure 9 shows CMOS output when All one's Applied.

Table 2. 90 nm technology.

Device	Output max value	Out min value	Avg. power	Delay	Static power		PDP
					$V_{IN} = 0$ V	$V_{IN} = 1.1$ V	
PMOS	1.2 V	0.233 V	13.26×10^{-9}	5 ps	1.3×10^{-15}	774×10^{-27}	663×10^{-20}
DTMOS	0.618 V	0.142 V	10.59×10^{-3}	6 ps	275.7×10^{-45}	20.28×10^{-3}	6.35×10^{-14}
NOVEL PMOS	1.2 V	0.158 V	10 nw	3 ns	15.60×10^{-15}	3.052×10^{-12}	3×10^{-17}
NMOS	1.08 V	0 V	8.245×10^{-9}	2 ns	1.404×10^{-24}	1.306×10^{-15}	1.6×10^{-17}
DTMOS	1.15 V	0.55 V	17.18×10^{-3}	4 ns	37.32×10^{-3}	959.1×10^{-30}	6.87×10^{-11}
PROPOSED NMOS	1.16 V	0 V	5.39×10^{-9}	0.0092 ns	45.24×10^{-12}	3.12×10^{-15}	4.3×10^{-21}

Table 3. 180 nm technology

Device	Output max value	Out min value	Avg. power	Delay	Static power		PDP
					$V_{IN} = 0$ V	$V_{IN} = 1.1$ V	
PMOS	1.8 V	0.674	1.086×10^{-6}	0.1215 ns	23.46×10^{-15}	32.74×10^{-30}	2.25×10^{-6}
DTMOS	0.6182	0.2942	991.40×10^{-3}	40 ps	0	1.8×10^{-3}	3.96×10^{-11}
NOVEL PMOS	1.8 V	0.450 V	1.08×10^{-6}	10 ps	46.96×10^{-15}	23.46×10^{-15}	108×10^{-17}
NMOS	1.22 V	0 V	0.398×10^{-6}	5×10^{-3}	522.8×10^{-48}	467.7×10^{-18}	199×10^{-9}
DTMOS	1.54 V	1 V	37.69×10^{-3}	10 ns	81.7×10^{-3}	106.2×10^{-30}	3.769×10^{-10}
PROPOSED NMOS	1.435 V	0 V	1.08×10^{-6}	6 ps	487.7×10^{-18}	$935.3 \times ^{10-18}$	648×10^{-18}

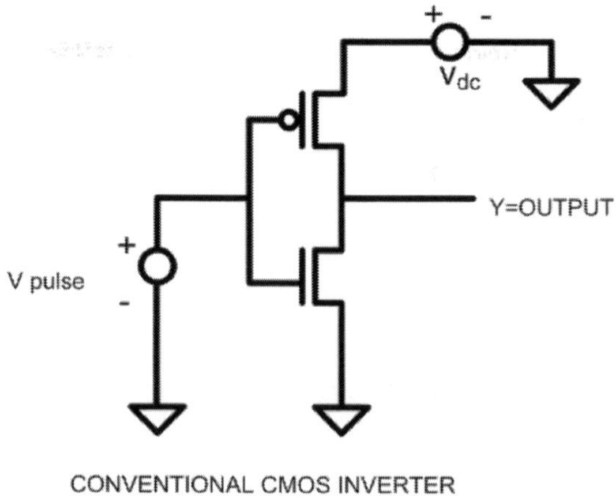

Figure 6. Conventional CMOS inverter.

INVERTER WITH PROPOSED PMOS AND NMOS

From the Figure 10 for T_6 and T_3 applied a reference voltage of 0.8 V when $V_{in} = 0$ V applied $T_4$and T_5 become on and output node will be 1 and T_1, T_2 will become off. When input is 1 T_4, T_5 will be non conducting and T_2, T_1 will become conducting will be zero as current sinks towards 0 v [8]. And output results are shown in the Figures 11-13.

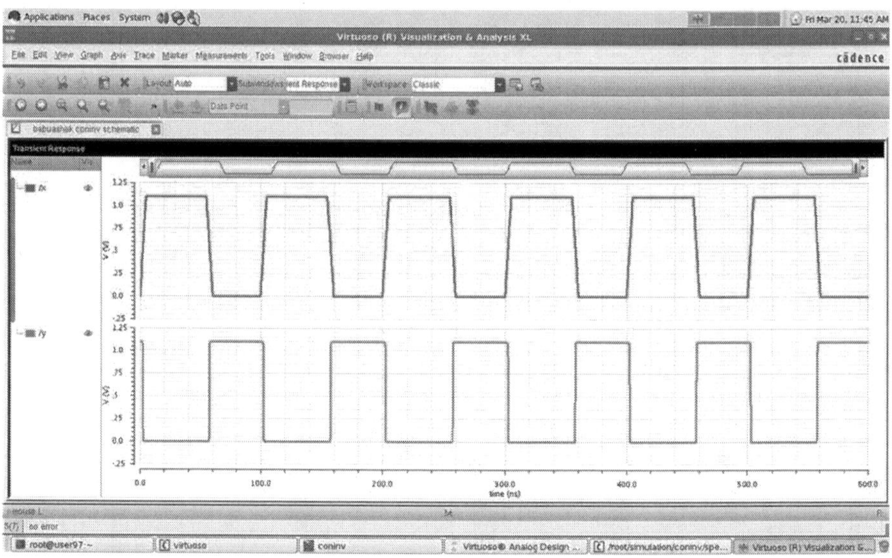

Figure 7. Conventional CMOS inverter output.

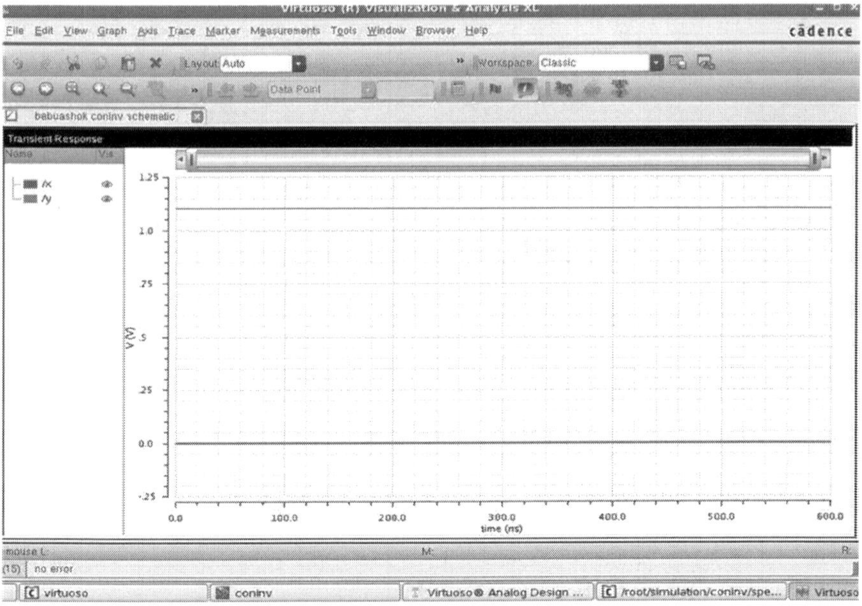

Figure 8. CMOS output when all zero's applied.

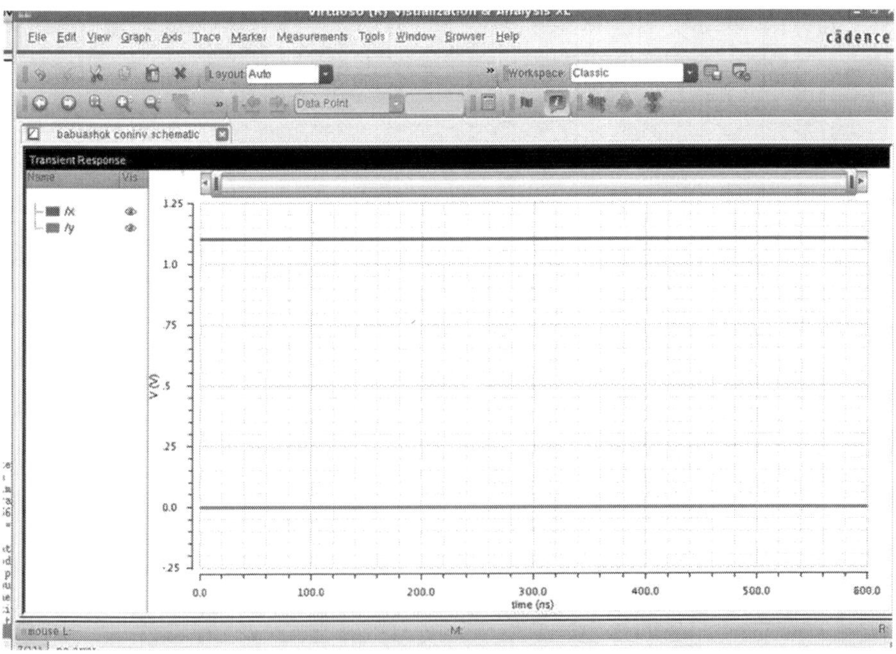

Figure 9. CMOS output when all one's applied.

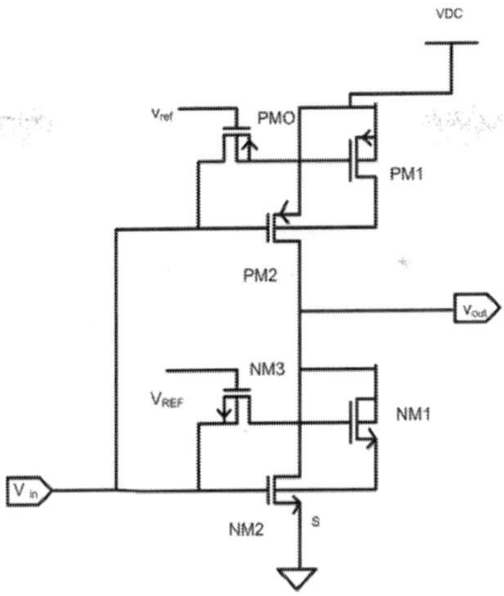

Figure 10. Proposed CMOS inverter.

There may be an area overhead but this circuit is certainly suitable for high speed and low leakage in deep submicron technology, as area is not a problem since we are in a position to keep millions of transistors in single die, performance parameters like speed will increase and leakage will decrease [9].

Hence an attempt has been made with proposed MOS devices to improve the performances.

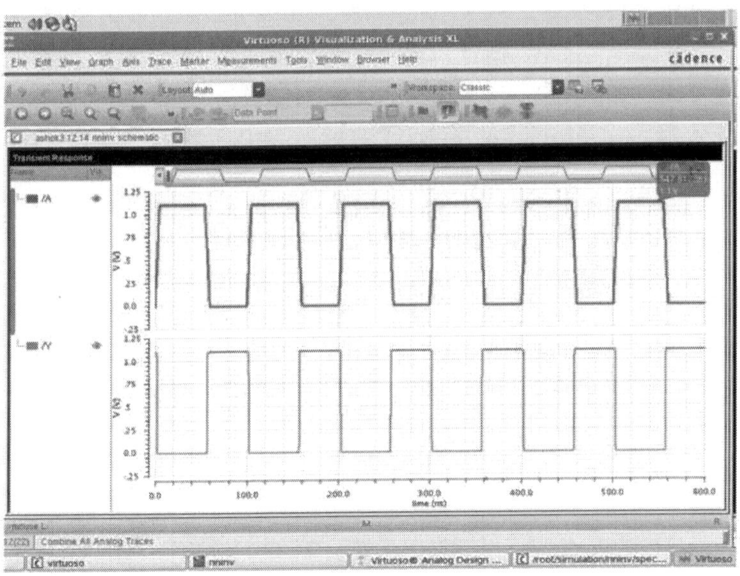

Figure 11. Proposed CMOS inverter output.

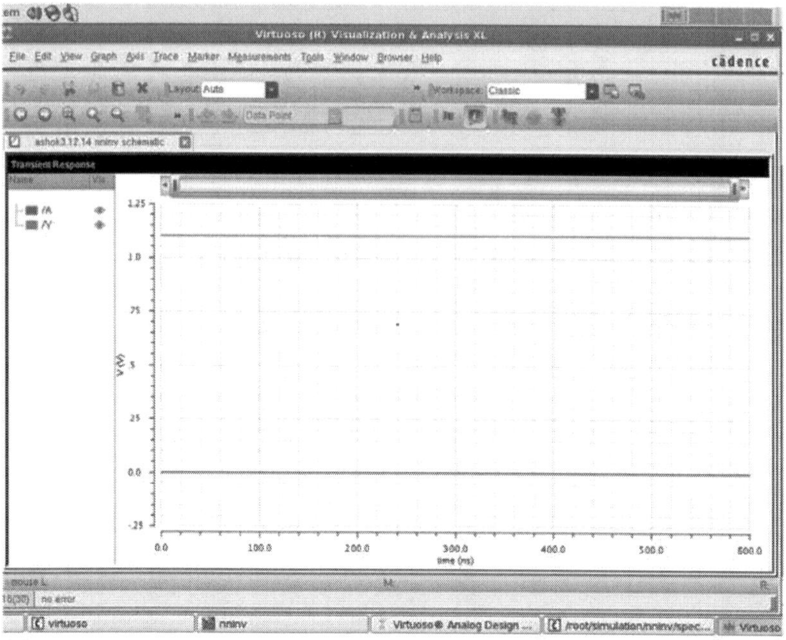

Figure 12. Proposed CMOS output for all zero's.

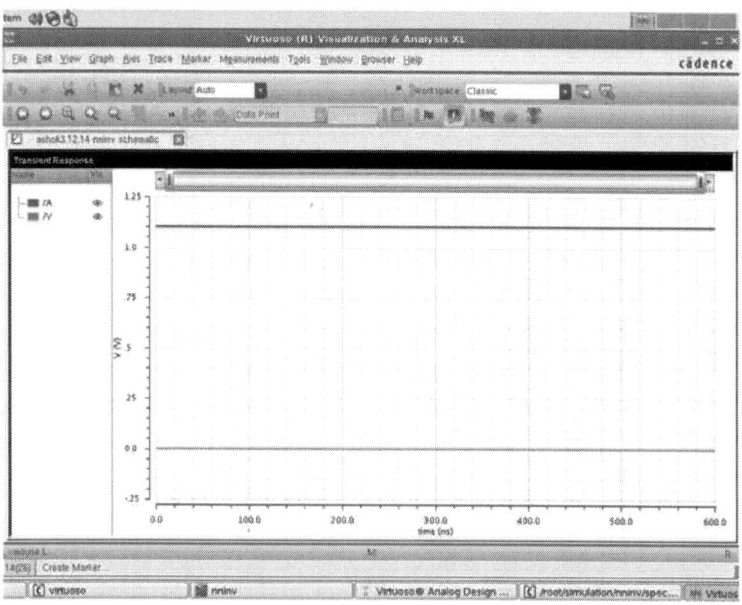

Figure 13. Proposed CMOS output for all one's.

LEAKAGE CURRENT MECHANISM

Reverse biased junction leakage and subthreshold leakage have very much similar characteristics both will be in order of pico amperes per device and very sensitive to process variations. Leakage current cannot be ignored in deep submicron region and leakage current is the beyond the digital designer control [10]. In large scale high performance digital chips performing operations with high frequency. In ultra submicron region leakage power is playing a major contribution apart of dynamic power [11].

Subthreshold channel leakage: the subthreshold conduction current is given by $I_{sub} = I_0 e^{v_{gs} - V_t / \alpha v_{th}}$ where alpha is parameter depends on fabrication process v_{th} is the threshold voltage, v_{th} is the thermal voltage, even though a transistor is logically turned off there may be non zero leakage current through the channel this current is known as Subthreshold leakage current Since it occurs below threshold voltage [12].

CALCULATION OF LEAKAGE POWER IN CMOS INVERTER AND PROPOSED CMOS INVERTER AT 45 NM TECHNOLOGY

Normal CMOS Inverter Leakage Power

- Power: 6.937 ′ 10^{-9}, delay: 0.05 ′ 10^{-9}.
- When all ones applied power is 4.917 ′ 10^{-12}.
- When all zeros applied power is 5.018 ′ 10^{-12}.
- Total power for all ones and all zeros is 4.97 ′ 10^{-12} + 5.018 ′ 10^{-12} = 9.935 ′ 10^{-12}.
- Leakage power = 6.937 ′ 10^{-9} – 9.935 ′ 10^{-12} = 6927 ′ 10^{-12}.

Novel CMOS Inverter Leakage Power

- AVG power P = 6.849 ′ 10^{-9}, delay = 0.18 ns.
- Power when all ones applied = 6.655 ′ 10^{-12}.
- Power when all zeros applied = 11.44 ′ 10^{-12}.
- All ones + all zeros power = 6.655 ′ 10^{-12} + 11.44 ′ 10^{-12} = 1.8095 ′ 10^{-11}.

- Leakage power = $6.849 ' 10^{-9} - 1.8095 ' 10^{-11} = 6830 ' 10^{-12}$ w.
- The difference of leakage between conventional CMOS and Proposed CMOS power is
- $6927 ' 10^{-12} - 6830 ' 10^{-12} = 97 ' 10^{-12}$.

CONCLUSION

We designed a Novel PMOS and NMOS which could give superior performance in deep submicron technology. The MOS devices like PMOS, DTMOS, and Novel MOS are studied and PDP is comparatively calculated. The leakage power is calculated for normal CMOS inverter and Novel CMOS inverter and it is observed that Novel CMOS inverter has less leakage power compared to Normal CMOS inverter. There may be area overhead but performance is better. It is concluded that Novel PMOS and NMOS are certainly suitable for low leakage and low power in ultra-deep submicron technology [13].

REFERENCES

1. Burch, R., Najm, F., Yang, P. and Trick, T. (1993) A Monte Carlo Approach for Power Estimation. IEEE Transactions on Very Large Scale Integration (VLSI) Systems, 1, 63-71.http://dx.doi.org/10.1109/92.219908
2. Ghosh, S. Devadas, K Keutzer, and J. White, (1992) Estimation of Average Switching Activity in Combinational and Sequential Circuits. ACM IEEE Design Automation Conference, 253-259.
3. Chou, T.L. and Roy, K. Statistical Estimation of Digital Circuit Activity Considering Uncertainty of Gate Delays. IEICE (Japan) Transactions on Fundamentals of Electronics, Communications and Computer Sciences, Accepted for Publication.
4. Chou, T.-L. and Roy, K. (1996) Accurate Estimation of Power Dissipation in CMOS Sequential Circuits. IEEE Transactions on VLSI Systems, 369-380.
5. Kumar, M., Hussain, Md.A. and Paul, S.K. (2013) An Improved SOI CMOS Technology Based Circuit Technique for Effective Reduction of Standby Subthreshold Leakage. Circuits and Systems, 4, 431-437. http://dx.doi.org/10.4236/cs.2013.46056
6. Kado, Y. (1997) The Potential of Ultrathin-Film SOI Devices for Low-Power and High-Speed Applications. IEICE Transactions on Electronics, E80-C, 443-454.
7. Cristoloveanu, S. and Reichert, G. (1998) Recent Advances in SOI Materials and Device Technologies for High Temperature. Proceedings of High-Temperature Electronic Materials, Devices and Sensors, San Diego, 22-27 February 1998, 86-93.

8. Brglez, F. and Fujiwara, H. (1985) A Neutral Netlist of 10 Combinational Benchmark Circuits and a Target Translator in Fortran. IEEE International Symposium on Circuits and Systems.
9. Chew, E.S., Phyu, M.W. and Goh, W.L. (2009) Ultra Low-Power Full-Adder for Biomedical Applications. IEEE International Conference of Electron Devices and Solid-State Circuits, Xi'an, 25-27 December 2009, 115-118.
10. Shalem, R., John, E. and John, L.K. (1999) A Novel Low Power Energy Recovery Full Adder Cell.
11. Chen, Z., Roy, K. and Chou, T.-L. (1997) Sensitivity of Power Dissipation to Uncertainties in Primary Input Specification. IEEE Custom Integrated Circuits Conference, 487-490.
12. Chen, Z., Roy, K. and Chou, T.-L. (1997) Power Sensitivity—A New Method to Estimate Power Considering Uncertain Specifications of Primary Inputs. ACM/IEEE International Conference on Computer-Aided Design, 40-44.
13. Najm, F. (1994) A Survey of Power Estimation Techniques in VLSI Circuits. IEEE Transactions on Very Large Scale Integration (VLSI) Systems, 2, 446-455.

CITATION

Ch, A. , Ravindra, J. and Lalkishore, K. (2015) Design of Ultra-Low Power PMOS and NMOS for Nano Scale VLSI Circuits. Circuits and Systems, 6, 60-69. doi: 10.4236/cs.2015.63007.

CHAPTER 2

Transaction-Level Power Analysis of VLSI Digital Systems

G.B. Vece, M. Conti and S. Orcioni

Department of Information Engineering, Università Politecnica delle Marche, Ancona, Italy

ABSTRACT

The increasing complexity of VLSI digital systems has dramatically supported system-level representations in modeling and design activities. This evolution makes often necessary a compliant rearrangement of the modalities followed in validation and analysis tasks, as in the case of power performances estimation.

Nowadays, transaction-level paradigms are having a wider and wider consideration in the research on electronic system-level design techniques. With regard to the available modeling resources, the most relevant framework is probably the transaction-level extension of the SystemC language (SystemC/TLM), which therefore represents the best platform for defining transaction-level design techniques.
In this paper we present a macro-modeling power estimation methodology valid for SystemC/TLM prototypes and of general applicability. The present discussion illustrates the implementation modalities of the proposed approach, verifying its effectiveness through a comparison with RTL estimation techniques.

INTRODUCTION

Over the last decade, several researches have been conducted to develop electronic system level (ESL) approaches in the modeling and analysis of microelectronic digital systems. The main motivations behind these researches come from the need of design techniques at an abstraction level higher than register transfer level (RTL). In fact, RTL techniques are becoming less and less suitable to face the complexity reached by many microelectronic products, as in the case of VLSI devices made up of million of gates. More precisely, the evaluations conducted on RTL prototypes may be onerous in terms of execution times and design efforts, because it is necessary to work with rather detailed descriptions of the system architecture. This can burden the design activities and make it difficult to meet strict time-to-market constraints. On the other hand, ESL design techniques allow to work on system prototypes derived from functional specifications, without having to deal with specific implementation details. In this way, there are more chances to take major decisions in the early design phases and with limited efforts, thus to speed up the flow towards the silicon implementation.

Transaction level modeling (TLM) has established itself as an effective ESL paradigm, especially in the description of the communication tasks between interacting modules. More precisely, when describing a complex digital system composed by several interconnected modules, a transaction-level representation allows a good separation between communication and elaboration tasks. The channel access is typically mediated by calls to interface functions capable to transport consistent data amount; each module is independent from the others with regard to interconnectivity [1]. The main benefits due to these features consist in the use of system prototypes that can be defined and simulated in reduced times. Moreover, the high interoperability of transaction-level models facilitates the realization of complex system architectures.

Nowadays, power performances are often a crucial constraint in the design of VLSI digital systems, in consequence of the wide diffusion of battery-supplied devices as well as the reliability issues due to high clock frequencies. The estimation of power dissipation is by now a primary design matter, providing useful indications in the analysis of implementation options and low-power solutions. Power estimation on RTL representations has been extensively studied, and a number of effective and well-tested techniques is now available [2], [3], [4], [5], [6], [7], [8] and [9]. More specifically, we can distinguish between cumulative and cycle-accurate RTL approaches [2]. In the first case, the power estimation is achieved by evaluating a power model at the end of a simulation period, on the basis of average input statistics. On the other hand, a cycle-accurate technique leads to evaluate the power model at every time cycle, requiring as input data cycle-based quantities.

In the design of a complex VLSI system, the realistic application of these estimation techniques is usually based on a modular approach. More precisely, the power model definition does not concern the system as a whole, but rather is carried out on basic modules that cover specific tasks. The RTL representation of the whole system is built through an aggregation of the RTL description of these modules; these latter can be realized by the designer or also taken from a third part library. In order to define the power model of a basic module, it is often necessary a characterization procedure which provides precise power consumption data from low-level power simulations [2]. For a specific CMOS technology, such characterization may be carried out once and for all, and the results may be reused for several VLSI systems based on that technology. Finally, once built the power models of the basic modules, we can perform power estimations on the whole system in its RTL representation.

RTL techniques are often capable to provide reliable power estimations, which can be used to verify design constraints as well as evaluate different implementation options. However, for the reasons before mentioned, many research interests are becoming more oriented towards analysis techniques at electronic system level. In particular, power estimations on transaction-level representations constitute an attractive investigation field that needs to be explored more deeply. In concrete terms, this means defining general and effective estimation methodologies applicable on transaction-level system descriptions.

From a conceptual viewpoint, we could expect a possible penalty in estimation accuracy when moving from RTL to transaction level power estimation techniques. In fact transaction-level techniques are applied on system prototypes at a higher abstraction level and with less implementation details than RTL representations. As a consequence, the run-time information achievable from a transaction-level prototype, and used in power model application, can be less detailed and precise. However, this possible loss of accuracy can be strongly mitigated if we consider a transaction-level estimation technique based on a macro-model approach [2], in which transaction-level metrics are connected to precise power measures coming from a low-level system implementation, such as the logic-level or the gate-level architecture.

The starting point for a research on transaction-level power estimation is to fix a reference language suitable for transaction-level modeling. In recent times, SystemC language has provided an extension specifically dedicated to transaction-level descriptions, (SystemC/TLM) [10], which is probably

the most relevant contribution in the formalization of a transaction-level modeling standard. As a consequence, SystemC/TLM is the most proper platform to define methodologies and CAD instruments for transaction-level power analysis.
SystemC language has already been at the center of some studies on transaction-level power analysis [11], [12], [13], [14] and [15]. However, some of the resultant techniques [11], [12] and [13] are referred to the core capabilities of the language, without applying the specific constructs provided by SystemC/TLM. In this way, they are precluded from the design advantages achievable from SystemC/TLM in terms of modeling facilities and simulation speed up. Examples of techniques relying on SystemC/TLM constructs are reported in [14] and [15]. The first contribution is a methodology based on the classification and power characterization of the functional tasks involved in transaction-level operations. The procedural steps are illustrated through a case study on IBM CoreConnect architectures. The experimental results show a good accuracy if there is a good matching between the considered functional tasks and the transactions generated in the run-time behavior of the system. This technique is applicable for SystemC/TLM prototypes in which the inter-module communications are based on the blocking transport interface [10]. On the other hand, the approach presented in [15] is a power state estimation technique aimed at evaluating the power dissipation in the main operative conditions and in reference to different system architectures. Even though the estimation results are not very accurate, the primary intent is to realize a fast comparison of the possible architecture alternatives. Analogously to [14], the application of this technique is restricted to prototypes based on the blocking transport interface.

This paper presents a transaction-level power estimation methodology valid for SystemC/TLM descriptions and of general applicability. In our researches, we have considered how to formalize the mapping of functional tasks onto the basic SystemC/TLM constructs that model transactions, without introducing restrictions related to the applied transport interface. By following an ad-hoc macro modeling technique [2] these SystemC/TLM constructs can be properly associated to energy estimations derived from measures on a low-level system implementation. In this way, we can estimate the energy of the transactions executed during a simulation session. Furthermore, we have studied the possibility to tune the estimation accuracy on the basis of the data transported by transactions.

In order to implement and verify the proposed approach, we have also realized a specific CAD tool that extends the SystemC/TLM language and is incorporated within the Power Kernel tool (PKtool) framework [13], [16], [17], [18], [19] and [20]. PKtool is a simulation environment born from an academic project and dedicated to power analysis on digital systems modeled in SystemC/C++. An open source release of the tool we have realized is available in [16].

This paper is structured as follows. In Section 2, a preliminary introduction on SystemC/TLM language is reported. Section 3 describes the details of the proposed methodology in conceptual and operative terms, while Section 4 provides the guidelines for a concrete application on SystemC/TLM prototypes. Section 5 explains how our methodology can be extended to cover power optimization techniques. Finally, some experimental results are illustrated in Section 6, considering also performance comparisons with RTL representations. Our methodology provides estimations in terms of energy; the ratio with simulation times allows to estimate also average power dissipations. To avoid ambiguities, in the remainder of the paper we often use expressions such as 'power dissipation' or 'power estimation' in conventional way, without an explicit reference to a physical quantity.

MODELING PRINCIPLES OF SYSTEMC/TLM

SystemC/TLM has been developed as an extension of the SystemC core language, with the aim to optimize the semantics and the simulation capabilities for transaction-level representations. Furthermore, the use of SystemC/TLM is regulated by precise guidelines that formalize a transaction-level modeling style with reference to specific detail levels.

In a typical RTL representation the inter-module communi-cations are realized through wire-like connections, pin-accurate I/O ports and clock-synchronized data transfers. Such scenario changes dramatically when moving to a SystemC/TLM description, in which the inter-module connections are virtually replaced by transport interfaces and the time references are restricted to the duration of transaction phases. Transaction-level communications are based on specific functions (TLM functions) that define the transport interfaces. More precisely, a transaction is modeled by means of calls to TLM functions that execute consistent data exchange between an initiator and a target module. The function calls may proceed from initiator to target (forward path) or in the reverse way (backward path). The linkage between an initiator and a target may be direct as well as constituted by intermediate modules that receive and forward TLM function calls (Fig. 1). In alternative mode, the modules involved in a TLM function call may be identified as caller and callee. The caller issues the function call, whereas the callee deals with the function execution according to a customized definition. The distinction between caller and callee is independent from the logical roles covered by a module as initiator, target or interconnect unit. In general, within a succession of TLM function calls, a module may be caller or callee in interleaved manner. The data exchanged in a transaction are typically referred to the format of a digital communication protocol. For this purpose, SystemC/TLM makes available a protocol template, called generic payload, suitable for memory-mapped bus (MMB) protocols and conceived to guarantee a high interoperability among models coming from different parts. However, generic payload includes some flexibility elements for a possible adaptation to specific communication protocols.

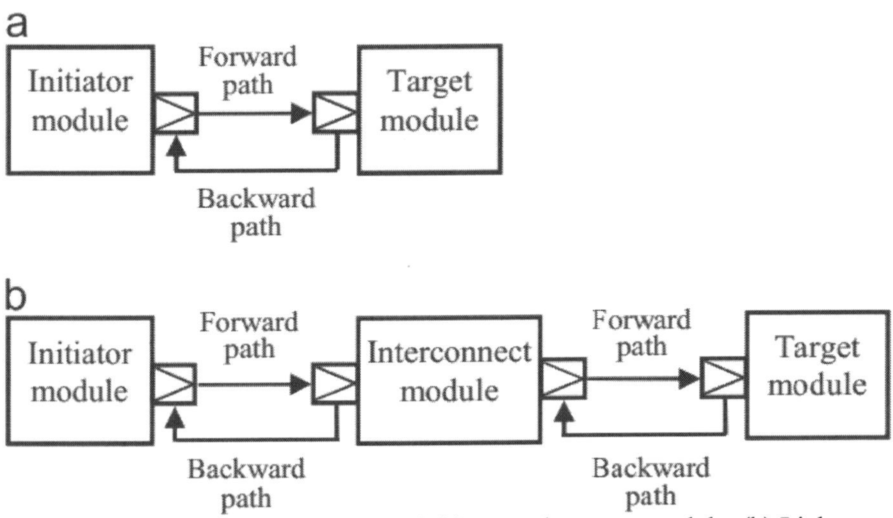

Figure 1. (a) Direct linkage between an initiator and a target module. (b) Linkage through an interconnect module (e.g. a router or an arbiter unit).

The TLM functions currently present in the SystemC/TLM framework are mentioned inFig. 2, with reference to their prototypes. The first three functions handle transactions in general manner, on the basis of blocking and non-blocking transport modalities [10]. In the former case, a transaction is entirely developed via a single function call on the forward path, having as unique time reference the overall processing latency. Conversely, the non-blocking modality allows a more precise modeling, with the inclusion of intermediate phases and timing points. In this way, the course of a transaction may be broken into more TLM function calls on the forward and backward paths. The functions get_direct_memory and invalidate_memory are dedicated to optimized operations for memory management. Finally, transport_dbg is reserved for debug purposes.

```
void b_transport
        (TRANS& trans, sc_time& t)

tlm_sync_enum nb_transport_fw
        ( TRANS& trans, PHASE& phase, sc_time& t )

tlm_sync_enum nb_transport_bw
        ( TRANS& trans, PHASE& phase, sc_time& t)

bool get_direct_mem_ptr
        (TRANS& trans, tlm_dmi& dmi_data)

void invalidate_direct_mem_ptr
        (uint64 start_range, uint64 end_range)

unsigned int transport_dbg
        (TRANS& trans)
```

Figure. 2. TLM functions provided by SystemC/TLM.

An important point is that TLM functions are not merely aimed at the transport of transaction data, but also carry out the elaboration of these data. The results of such elaborations are reported in the return value and/or in the references passed as input parameters. Actually, TLM functions may be regarded as the primary elements to represent the run-time functionality of a module in SystemC/TLM. In modeling terms, this is realized by the mapping onto two main entities: the bodies of TLM functions and the processes that issue TLM function calls (Fig. 3). In practice, the whole dynamic behavior can be reconducted to these two entities, because on them depend the tasks related to I/O evolution. In regard to power estimation, this means that dynamic power dissipation can be evaluated by monitoring TLM function executions and caller processes.

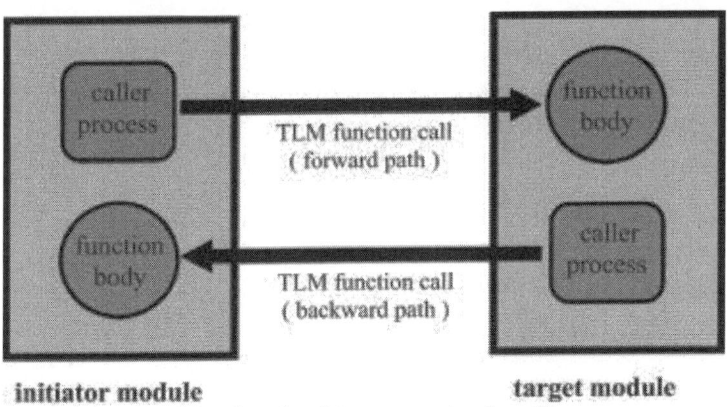

Figure 3. Functional entities involved in transaction handling.

DESCRIPTION OF THE PROPOSED METHODOLOGY

Power Estimation of System Modules

To illustrate the proposed methodology, the starting point is to characterize properly the power dissipation of a transaction- level prototype defined in SystemC/TLM. In general, the occurrence of a transaction entails an energy dissipation for all the involved modules (the initiator, the target and possible interconnect units). Consequently, the dynamic energy dissipation of a single module can be achieved considering the energy contributions of the transactions in which it is involved. In this way, the overall energy dissipation can be estimated as the sum of the estimations due to the transactions occurred during a simulation.

The realization of this approach requires to sample the calls to the TLM functions constituting a transaction and to estimate an energy dissipation for each module involved in such function calls. For this purpose, each module should be associated to a set of power models, one for each TLM function that may concern the module in the role of caller or callee. In this context, a power model can be regarded as a formal definition of the energy cost of a TLM function, from the caller or the callee side. Every time a TLM function is issued, an energy estimation is computed for the caller and callee modules, by applying the respective power models linked to such function (Fig. 4).

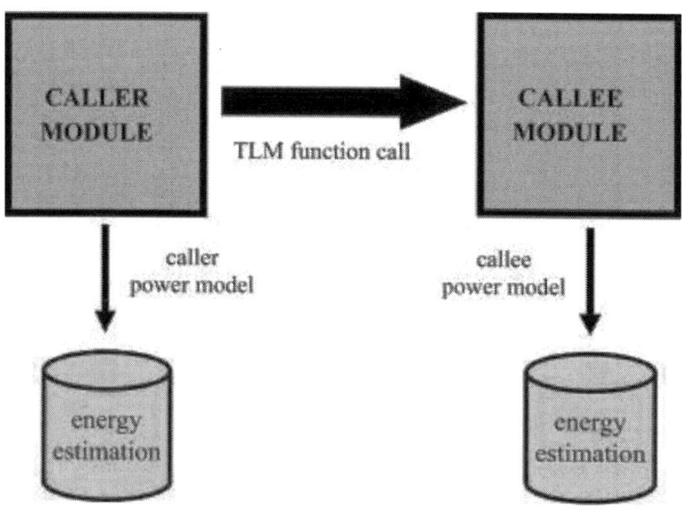

Figure 4. Energy estimations computed in a TLM function call for the caller and the callee module.

An essential aspect is to understand how much this estimation modality fits the effective power dissipation of a module. For this purpose, we can evaluate the coverage level with respect to the module functionality, mapped in this case on the TLM function bodies and the caller processes (Fig. 3). The power dissipation of the TLM function bodies is completely covered by the function calls from the callee side. As concerns the caller processes, the related dissipation can be achieved from TLM function calls through an extended view of their energy costs. More precisely, in the energy cost of a TLM function from the caller side, it is always possible to include the contributions of the elaborations surrounding the function call instruction inside the process. In principle, this energy cost could include the whole execution flow of the caller process. If a process issues distinct TLM function calls in different points of its execution flow, the energy cost of each function could include only the closest elaborations, in an orthogonal manner and up to cover the whole execution flow. From these observations we can conclude that the considered estimation approach is able to cover the whole functionality of a module, and provide non-partial estimations in regard to the dynamic behavior.

As concerns the power model extraction (that is how to determine the energy costs of a TLM function with respect to the caller or callee side), we can apply a specific macro-modeling technique [2]. The first step consists in a characterization procedure aimed at determining accurate measures of the energy dissipation of the function. These measures should be achieved from power simulations on a low-level prototype of the examined system, e.g. a gate-level description. More precisely, the low-level prototype should be simulated to reproduce the execution of single function calls on a significant amount of training input stimuli. From each simulation, by evaluating an accurate gate-level power model, we obtain the energy dissipation of the TLM function for the applied input sequences. Finally, by averaging all the obtained measures, we can achieve the energy costs that define the power model associated to the TLM function. There are no specific limitations on the power model that can be used in this procedure. It is possible to use a power model based only on dynamic power and signal activity as well as more accurate power models including also leakage power contributions or any other source of power dissipation.

Entering more into details, this characterization procedure requires to map the SystemC/TLM prototype on the low-level prototype, in order to distinguish those circuitry blocks triggered by the TLM function call (Fig. 5). In low-level power simulations the sum of the energy dissipations of these circuitry blocks will provide the energy measures related to the TLM function call.

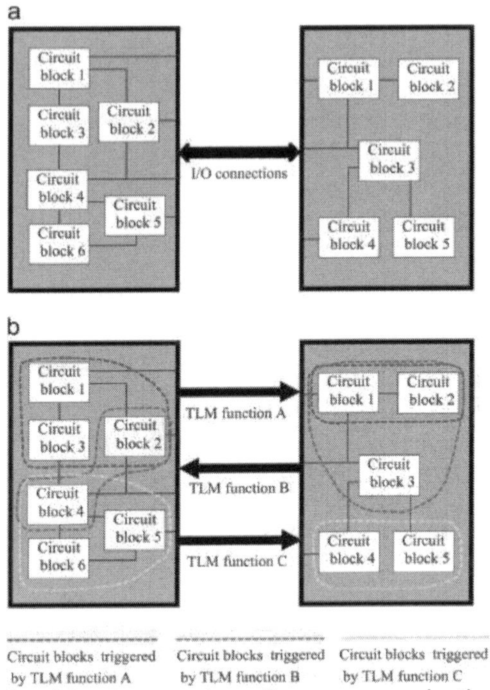

Figure 5. (a) low-level implementation of two communicating modules through the constituent circuit blocks. (b) Mapping between the constituent circuit blocks and distinct TLM functions.

As concerns the input stimuli applied in low-level simulations, it is important to consider the time duration associated to the TLM function call. More precisely, the data exchanged in a transaction could cover several clock cycles in the low-level prototype, spanning a sequence of consecutive input values (and not simply a pair of these latter). For this reason, the training stimuli set should consist in a significant number of input sequences corresponding to typical data exchanged in transactions.

In our methodology the power model is referred to a table-based representation [2], in which the energy costs of a TLM function call are stored in a lookup table for the caller and the callee module. In its simplest version, the characterization procedure leads to get two generic energy costs, one for the caller and another for the callee side, used as estimations every time the TLM function is run. In this case, the power model is merely associated to lookup tables with only one element. Nonetheless, more accurate power models may be obtained considering also the I/O information managed by the TLM function, i.e. the input parameters and the return value (TLM data). Such information are strictly related to the current operative conditions, such as the type of operation, the transaction phase, the transaction status. The effective power dissipation may significantly vary for different TLM data entries, because of their influence on the signal dynamics and the elaborations carried out. To clarify this point, we can consider the examples reported in Fig. 6, which show two typical SystemC/TLM code fragments representing a caller process and the body of a TLM function. In the first example, a TLM function is called with the subsequent execution flow depending on the value returned by the function; this return value specifies the status of the transaction. Examining the code, it is evident how different return values cause the execution of different elaborations, which in turn may lead to quite different power dissipations. The same conclusions hold also for the other example, in which the power dissipation depends on the operation type associated to the transaction.

a

```
tlm_sync_enum  status =
            initiator_socket>nb_transport_fw
                (*transaction_ptr, phase, delay);

switch (status)
   {
     case tlm::TLM_ACCEPTED:
     {
       //execution flow #1
       ...
     }
     case tlm::TLM_UPDATED:
     {
       //execution flow #2
       ...
     }

     case tlm::TLM_COMPLETED:
     {
       //execution flow #3
       ...
     }
   }
...
```

b

```
void b_transport
    (transaction_type& trans,sc_time &t)
{

  if (trans.get_command() == TLM_WRITE_COMMAND)
  {
    // 'write' execution flow
    ...
  }
  else if(trans.get_command() ==
                                 TLM_READ_COMMAND)
  {
    // 'read' execution flow
    ...
  }
  ...
```

Figure 6. Examples of execution flows depending on TLM data in a caller process (a) and a TLM function body (b).

TLM data can be taken into account to get more precise measures from the characterization procedure, thus improving the accuracy of the energy costs associated to a TLM function. This opportunity is based on the mapping of a transaction-level prototype onto a gate-level architecture. In fact, in such mapping several TLM data may correspond to signals of easy identification in the gate-level description. Focusing on one of the TLM data, for example the transaction status, we can realize distinct characterization procedures for fixed values of the corresponding gate level signals. In this case, we can carry out specific evaluations of the energy dissipated by the caller process, whose execution flow may depend on the transaction status specified by the return value of the TLM function. At the end of the characterization procedure, what we end up with is something more accurate than an overall energy cost, i.e. the average power dissipations related to the possible values returned by the TLM function. Each of these measures can be assumed as the energy cost of the function in the corresponding transaction status. In this way, the TLM function can be associated to several energy costs from the caller side, according to the cardinality of the transaction status. Accordingly, the power model is associated to lookup-tables with as many elements as the possible transaction status. During a simulation on the transaction-level prototype, in each function call, the energy cost of the caller side will be the one associated to the current transaction status, as specified by the return value of the function.

To explain better this characterization procedure, we can consider the low-level architecture in Fig. 7, which shows a general representation of the caller circuitry involved in a TLM function. In such architecture we can distinguish a combinatorial unit, a flip-flop array holding the internal state, the binary primary inputs (I1, I2,..., I*n*) and the binary state signals (S1, S2, ... S*m*). At this point we should consider the mapping between TLM data and the binary input/state signals. In particular, the inputs i2–i3 and the state signal s2 could be associated to the transaction status, which can assume three different values, i.e. TLM_ACCEPTED, TLM_UPDATED, TLM_COMPLETED [10]. In this example, we can suppose the following correspondence with the binary values of (i2,i3,s2): (0,1,0). (1,0,0), (0,0,1).

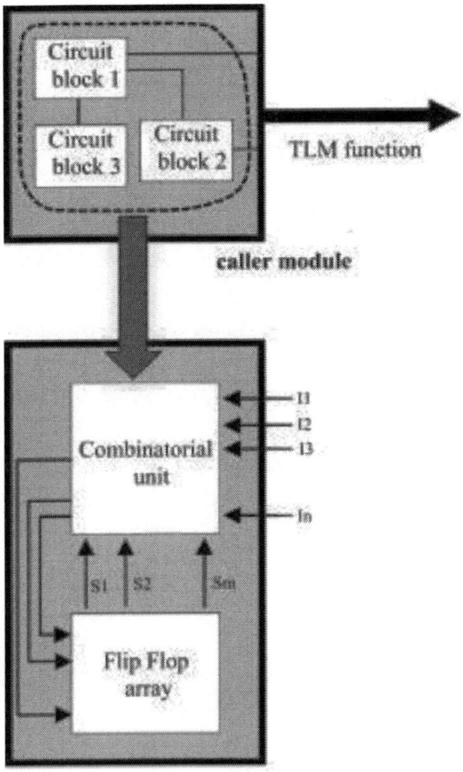

Figure 7. General low-level representation of the circuit blocks triggered by a TLM function.

During the characterization procedure, if we want to achieve energy measures specialized with respect to the transaction status, we should executed a number of power simulations where i2, i3 and s2 are fixed whereas the other signals are assigned to sequences of variable values. These power simulations should be repeated for the three possible transaction status, maintaining (i2,i3,s2) fixed to the corresponding binary values. By averaging the energy measures resulting from power simulations, we achieve three energy costs (one for each transaction status) which will be the elements of the lookuptable power model. These costs represent the estimations of the energy dissipated by the caller module when the considered TLM function is called. More specifically, in each function call the energy cost selected will be the one associated to the transaction status returned by the function.

The specialization process now described can be further refined, considering jointly more TLM data and deriving power models with more numerous and precise energy costs. These latter will be associated to the possible combinations of the TLM data values. Such a refinement may further improve the estimation accuracy at the price of a more articulated characterization procedure. In this way, the resulting power model will be represented by a multi-dimension lookup table, with as many dimensions as the number of considered TLM data.

The approach so far discussed is able to estimate properly only the power dissipation related to the dynamic behavior of a module. In fact, TLM functions reproduce only the active phases of the run-time evolution, without including those time periods in which no transaction is handled and the power dissipation is essentially due to leakage effects [21]. In general, the run-time behavior of a module is constituted in its entirety by an alternation between active and idle phases, as illustrated in Fig. 8. Especially in VLSI technologies with high integration levels, these idle phases may take a relevant fraction of the total power dissipation such that their effects could not be secondary.

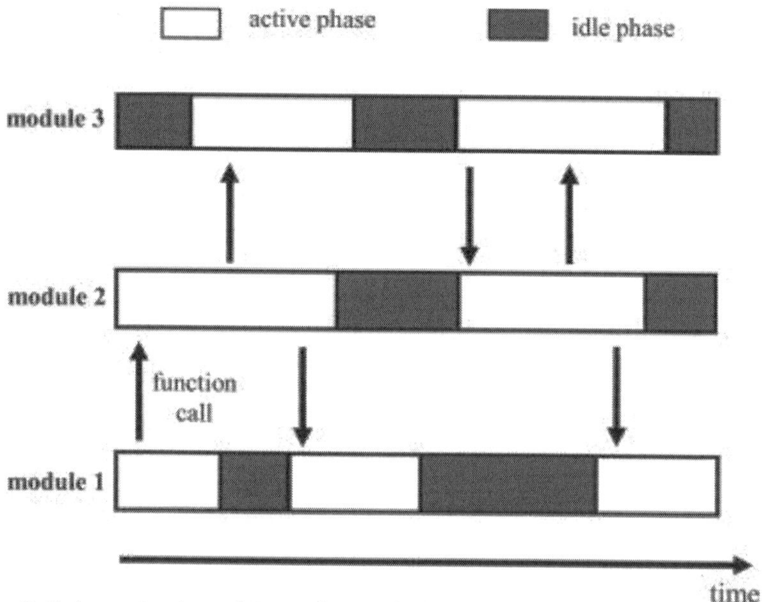

Figure 8. Schematization of the active and idle phases in the run-time evolution of a transaction-level system prototype.

To solve this limitation, it is possible to consider the solution proposed in [14], in which a further contribution is computed for including the power dissipation in idle phases. In analytical terms, this leads to the following estimation formula:

$$E_{mod} = \sum_{i}^{N} e_i + P \times T_{idle} \tag{1}$$

where *N* is the number of transactions occurred during a simulation, *ei* the energy of a single transaction (achieved from lookup-table power model), *Tidle* the sum of the idle time intervals and *P* the idle power dissipation. Eq. (1) represents the energy dissipation of a single module and *ei* is referred to the call of a TLM function that may concern the module in the role of caller or callee.

Power Estimation of Inter-Module Connections

This subsection briefly illustrates the application of the proposed methodology for including inter-module connections in power estimation. In modern VLSI systems the wire connections between modules can have a relevant impact, especially in FPGA-based implementations [21]. In our methodology it is possible to take into account also the power dissipation of wire connections. For this purpose it is necessary to extend the low-level characterization procedure described in the previously subsection.

Considering the wire connections between two interacting modules, we could determine the energy cost due to their power dissipation through the same low-level simulations run to calculate the energy costs of the TLM functions of the two modules. As power model, we could use the following expressions:

$$E_{wire} = \frac{1}{2} \sum_{i}^{N} S_i C_{avg} V_{DD}^2 \tag{2}$$

where *N* is the number of wires that connect the two modules, *Si* the signal activity of each wire, *Cavg* the wire average capacitance and *Vdd* the applied power supply.

In order to apply Eq. (2), it can be important to have an idea of the area occupation of the inter-module connections in the examined VLSI system; in fact this piece of information can be necessary to estimate the average wire capacitance.

Also for inter-module connections we could apply a TLM data specialization, i.e. the wire energy cost can be possibly refined in reference to entries of one or more TLM data, by following the same procedure described in the previous subsection. In this way, at the end of the characterization process, we would end up with a further lookuptable energy costs associated to inter-module connections.

APPLICATION GUIDELINES

After this conceptual description, it is important to consider how the proposed methodology can be concretely applied on a SystemC/TLM prototype. In particular, we should consider the application modalities on the single modules constituting the prototype. The best solution is probably to extend the basic implementation of the native SystemC/TLM modules, by adding the functionalities required for power estimation tasks. This approach can be addressed by leveraging on the inner implementation of the SystemC/TLM language, represented by an open-source C++ class hierarchy in which it is possible to modify the native classes used in module description. Given a SystemC/TLM prototype composed by several interconnected modules, without loss of generality, we can focus on two modules that are involved in a transaction execution. In this example, such modules are called module_a and module_b and play respectively the roles of caller and callee.

To apply our approach, the first intervention could be done on the TLM function bodies defined inside the module classes. These functions have been listed in Fig. 2 and implement the tasks for handling a transaction that may involve the two modules. Each of these TLM function is derived from an interface class and must be defined by the module that covers the callee role when the corresponding TLM function is called. More precisely, every time the TLM function is called by a process of the caller module, the TLM function body is executed. As a consequence, the energy contributions related to the transaction could be sampled and computed inside the TLM function body. In concrete terms, this means to override the interface function defined inside the calle class, so that such function carries out its basic task and, in addition, computes the energy contributions for the caller and callee module.

Coming back to our example, the pseudo-code in Fig. 9 shows this enhanced implementation in the case of the TLM function nb_transport_fw defined inside the module_b class. The functions nb_trfw_energy_caller and nb_trfw_energy_calle are aimed to get the energy estimations for the caller and the callee module involved in the transaction, by querying the lookup-table power models. These functions can be defined inside the calle class. The estimated energy contributions are summed up to trans_energy_module_a and trans_energy_module_b, two variables holding the total energy estimated during the simulation for module_a and module_b. Each of these variables could be defined in the class of the corresponding module. For this reason trans_energy_module_a is handled through a pointer referred to the module_a object. The functions nb_trfw_energy_caller and nb_trfw_energy_calle take as input parameter the TLM data related to the transaction, in order to select the correct energy cost from the lookup-table power model. nb_trfw_energy_caller is called after the original instructions and takes as input parameter also the return response of nb_transport_fw. This implementation is necessary because the values of some TLM data are known from the caller side only after the original TLM function is executed, as in the case of transaction status.

a

```
tlm::tlm_sync_enum
  module_b::nb_transport_fw
  ( tlm::tlm_generic_payload &gp,
    tlm::tlm_phase            &phase,
    sc_core::sc_time          &delay_time)
  {

   // instructions for transaction handling
   ...
   ...
   return status;

  }
```

b

```
tlm::tlm_sync_enum
  module_b::nb_transport_fw
  ( tlm::tlm_generic_payload &gp,
    tlm::tlm_phase            &phase,
    sc_core::sc_time          &delay_time)
  {

   trans_energy_module_b +=
         nb_trfw_energy_callee(gp, phase);

   // instructions for transaction handling
   ...
   ...

   mod_a->trans_energy_module_a +=
         nb_trfw_energy_caller(gp, phase,
                               return_status);

   return status;

  }
```

Figure 9. (a) Original implementation of nb_transport_fw in the callee class. (b) Addition of the instruction to compute energy estimations.

The lookup-tables with the energy costs of the TLM function may be created and linked to the caller and callee modules at the beginning of the simulation, by defining all the necessary instructions in the constructors of their classes. In particular, the energy costs defining each lookup-table could be explicitly assigned in these constructors or could be read by a text/xml file edited by the user. At software level the two lookup-table of the TLM function can be simply implemented through multi-dimension arrays of floats, where the number of dimensions corresponds to how many TLM data are considered to refine power estimations. In this way, the overall lookuptable size (i.e. the number of energy costs stored in the table) is given by the product of the entries of the considered TLM data. In computational terms, querying a C++ array is a quite-fast operation; as a consequence, during a SystemC/TLM simulation the power model application should not burden the simulation speed in relevant way. Actually, the most expensive operation in power model handling is probably the construction of the lookuptable at the beginning of the simulation. In particular, this means reading the energy costs from a text file and create the array of floats that store the energy costs. Nonetheless, the overhead of such operation can be strongly reduced for simulations conducted on relatively long observation periods.

The instructions to display the total estimation results (i.e. the variables trans_energy_module_a and trans_energy_module_b) could be reported in the destructors of the module classes. In compliance with the C++ execution rules, the destructor functions will be called at the end of the SystemC/TLM simulation, when the overall power estimations are available.

At this point, in compliance with eq. (1), the last contribution to be estimated is the static energy related to the idle periods. This means to estimate for each module the time period during which it is in an idle state and no transaction is executed. For this purpose, we can introduce a further variable in the classes of module_a and module_b, which is aimed to hold the total active time. During a simulation, likewise the transaction energy estimations, these variables could be updated inside the TLM function bodies, using the time parameter related to the transaction duration. Such parameter is passed as writeable reference to the TLM functions in both blocking and non-blocking transport interface. Accordingly, the pseudo-code shown in Fig. 9 should be revised according to the version in Fig. 10. For each module the total idle time could be finally computed in the class destructor, by subtracting the total active from the simulation time. After that, in the destructor we could report the instructions to display also the idle energy estimation.

```
tlm::tlm_sync_enum
module_b::nb_transport_fw
( tlm::tlm_generic_payload &gp,
  tlm::tlm_phase            &phase,
  sc_core::sc_time          &time)
{

 trans_energy_module_b +=
       nb_trfw_energy_callee(gp, phase);

 // instructions for transaction handling
 ...
 ...

 mod_a->trans_energy_module_a +=
       nb_trfw_energy_caller(gp, phase,
                             return_status);

 // active time updating
 active_time_module_b += time;
 mod_a->active_time_module_a += time;


 return status;

}
```

Figure 10. Addition of the instruction to compute active times in nb_transport_fw.

If we want to consider also the power estimation of the inter-module connections, we should simply add a further instruction in the pseudo-code in Fig. 9 and Fig. 10, which handles the energy dissipation due to the connection wires during the TLM function call. Analogously to the instructions seen for module_a and module_b, this further instruction should execute a specific function that returns the energy estimation of the connection wires by querying the lookup-table with the associated energy costs. The estimation returned by this function will be summed up to a dedicated variable holding the total energy estimated for the wire connections during the simulation.

By following these application guidelines, it possible to define all the functionality necessary to run our methodology on a SystemC/TLM description during an ordinary SystemC/TLM simulation. The implementation proposed in this section is simple, effective and may have a limited impact on the time performance of an ordinary SystemC/TLM simulation. On the basis of these guidelines we have realized a CAD tool for power simulations on SystemC/TLM prototypes, which provides user-friendly constructs to apply the proposed methodology. This tool extends the capabilities of the PKtool simulation environment and is currently available in [16].

REALIZATION OF POWER OPTIMIZATION TECHNIQUES

In modern VLSI systems there are often implemented techniques for optimizing power dissipation, such as dynamic voltage/frequency scaling (DVFS). An effective power estimation methodology should take into consideration such optimization techniques and model the related effects on the system performances.

A noteworthy study on DVFS modeling in SystemC/TLM prototypes is illustrated in [22]. In this research is introduced the concept of power state, as entity to distinguish the working periods in which the system frequency/voltage can assume fixed values. In this case the power state characterization is based on a functional-level mapping, in which the power states are associated to the different functional conditions which can occur in the run-time behavior.

Also in our methodology it is possible to model the contribution of DVFS techniques in power estimations. This modeling can be realized by connecting power states to TLM functions and determining the current power state when a TLM function is called. In comparison with [22], this approach provides a better formalization with reference to the modeling constructs of SystemC/TLM languages. This, in turn, can make easier and more systematic the description of DVFS tasks in power model application.

In concrete terms, the modeling of DVFS techniques can be implemented through an augmented version of the energy-cost lookup-table associated to a TLM function. That is to say, we should define as many lookup-tables as the possible power states in which the system can work. As an example, we can consider a dynamic voltage scaling scenario where the voltage can switch among three different levels, on the basis of the current operative state. In this scenario, for a fixed set of TLM data, the power dissipation of a TLM function can be associated to three different energy costs, one for each voltage level. These costs should be derived from three distinct sessions of low-level power simulations, in which we consider the different voltage levels on the gate-level prototype.

In this way, such approach leads to a power model lookup-table with a further input, in addition to the proper TLM data, i.e. the applied voltage. Considering the power model implementation described in Section 4, this implies to increase of one dimension the array storing the energy costs. With regard to power model characterization, a higher simulation effort could be required, because it may be necessary to repeat the low-level power simulation for all the possible voltage levels.

To complete the DVFS modeling, it is necessary to realize the state machine that determines the current power state, i.e. the voltage/frequency currently applied. This state machine should be incorporated inside each caller/callee module that relies on a DVFS power optimization. For this purpose, we could define an internal submodule that implements the rules to detect the current power state (Fig. 11). This submodule should have access to all the information required by these rules, in particular the input and internal state signals on which depend the operative conditions. Moreover, this submodule should be able to communicate the current power state to the lookup-table power model, in order to select the correct energy costs. In compliance with the implementation described in Section 4, the current power state can be reported in a variable shared with the two functions that returns the energy estimations of a TLM function call.

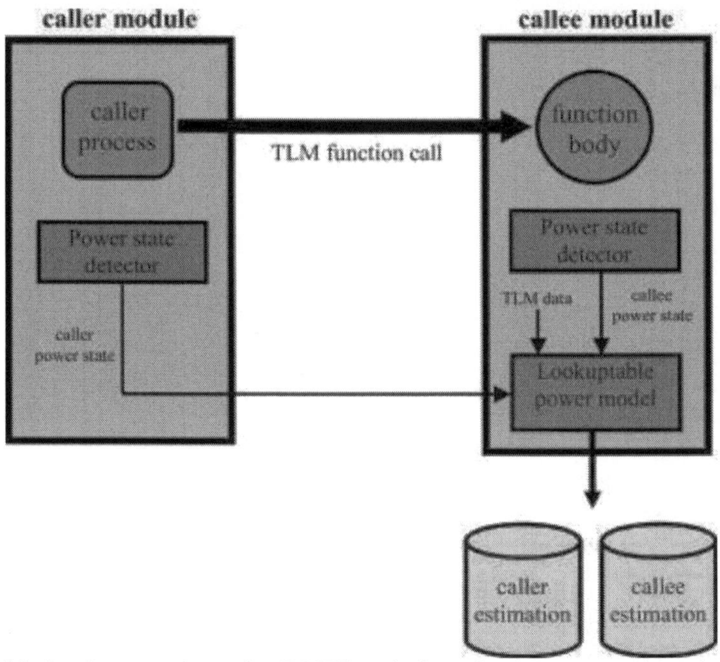

Figure 11. Implementation of a DVFS technique in the proposed methodology. Solution compliant with the application guidelines provided in Section 4.

EXPERIMENTAL RESULTS AND PERFORMANCE EVALUATION

Study on SystemC/TLM Demonstration Examples

To evaluate the performances of the proposed methodology, we have conducted specific simulations considering the comparison with RTL estimation techniques. For this purpose, we have used the simulation tool introduced in Section 1 (the PKtool framework) and, as testing platform, we have considered benchmark systems of variable complexity, taken from SystemC/TLM releases and SoCLib research project [23].

First of all we have considered three transaction-level prototypes included in SystemC/TLM releases (TLM-2.0.1); they are p2p/hierarchical_socket, at_1_phase and at_4_phase [10]. Such prototypes are fully compliant with the modeling rules specified in SystemC/TLM documentation, and represent demonstration examples given by the aggregation of typical digital components. p2p/hierarchical_socket reproduces a simple peer-to-peer connection between two modules implementing datapath operations. The communication channel is based on the blocking transport interface and each transaction is executed through a single TLM function call. at_1_phase and at_4_phase define more articulated bus-based systems, in which the inter-module communications rely on the non-blocking transport interface and a transaction may be carried out through more TLM function calls.

At the beginning of this study, we have realized in SystemC a gate-level description for each of the examined systems, using the fine-grained constructs of the language for bit operations. The resulting architectures are constituted by FSMD blocks and memory units whose implementation is based on elementary logic ports. From these gate level architectures we have firstly computed the energy costs of the TLM functions issued in the transaction-level prototypes. These costs have been determined considering a generic behavior (with execution flows randomly generated) and also considering operative conditions related to specific TLM data, i.e. transaction phase and status. Furthermore, we have used these gate-level prototypes to determine the power dissipation in typical running sessions. The resulting measures have been assumed as reference terms to evaluate the estimation accuracy.

Function costs and dissipation measures have been achieved from SPICE-like simulations, in which we have adopted a power model based on the expression for the energy dissipation of a CMOS cell [21]. The gate-level descriptions can cover transient assignments and spurious commutations, thus taking into account also glitch effects. As concerns technology mapping, we have considered the characteristic data of a 90nm process to set the power model parameters. The low-level power simulations have been executed by leveraging on the capabilities of the PKtool framework, which allows a run-time computation of the gate energy dissipation In particular, as concerns the dynamic part of power dissipation, during a simulation PKtool allows to sample the gate commutations by means of dedicated components called augmented signals. The data on commutations represent the fundamental information to evaluate the dynamic energy dissipated by the single gates, by applying the expression for the energy dissipation of a CMOS gate. In this way, by summing up the energy dissipation of all the single gates, we can achieve the energy dissipation of the examined system during a simulation.

In order to realize a more comprehensive study, we have described in SystemC also RTL prototypes of the examined systems. The comparison with these RTL counterparts has allowed to extend our analysis to other relevant design metrics, in addition to power dissipation, such as modeling effort and simulation speed. This comparison is validated by the use of the same language to describe and simulate the prototypes in the different abstraction levels. To run the SystemC simulation engine, it is only necessary a C++ compiler environment, e.g. Visual C++ or GNU g++.

As alternative solution, these low-level power simulations could be carried out on gate-level prototypes realized in a language more supported by standard CAD tools (e.g. VHDL language). In this case we would have probably achieved more precise energy costs. Nonetheless, the application of our methodology would have led to the same power estimation accuracy, because our methodology is based on transaction-level constructs that are independent from the gate-level implementation and the precision of the low-level energy measures.

Table 1 and Table 2 show the results of evaluations, in which we have determined the code size of the different prototypes and the CPU times required to simulate them. All the simulations have been executed on a PC unit with a 2500 Mhz dual core Intel processor and 4 GB of RAM.

Table 1. Code size of the SystemC/TLM examples in the different abstraction levels and TL-RTL effort reduction.

Abstraction level	p2p/hs	at_1_phase	at_4_phase
Gate level	113 KB	340 KB	370 KB
RTL	24 KB	206 KB	211 KB
TL	17 KB	126 KB	130 KB
Effort reduction (RTL−TL)/RTL×100	29%	39%	38%

Table 2. Speedup factors with respect to gate-level simulations and between TL-RTL simulations.

Abstraction level	p2p/hs	at_1_phase	at_4_phase
RTL	30x	310x	290x
TL	120x	3030x	3000x
TL/RTL	4x	9.8x	10.3x

As confirmed by these data, moving from RTL to transaction level representations entails evident improvements with reference to the considered metrics. The modeling efforts may be reduced in remarkable measure, even by 30–40% for prototypes based on the non-blocking transport interface. With regard to CPU times, the first two rows of Table 2specify the speedup factors with respect to gate-level simulations. The ratio between these values, reported in the bottommost row, provides the speed up of transaction-level simulations with respect to RTL ones. We can verify how transaction-level simulations are definitely faster, with a time saving that can reach one order of magnitude in the case of non-blocking transport modalities.

In the core part of our analysis, we have estimated the power dissipation of the examined systems by applying our technique on the transaction-level prototypes. For this purpose, we have considered three power model variants, i.e. model_0, model_1 and model_2. In model_0 a TLM function is associated to its generic energy cost, which is assumed as estimation in every function call without applying TLM data specialization. model_1 introduces a first-level refinement, comprising energy costs related to the possible entries of one TLM data. We have studied two cases for model_1, considering separately transaction phase and status as TLM data. Finally, in model_2 the energy cost of a TLM function call depends jointly on transaction phase and status. The analysis on p2p/hierarchical_socket has been restricted to model_0, since the blocking transport interface does not handle phase and status information.

Table 3 shows the average errors resulting from transaction-level estimations, with respect to the reference measures achieved from the gate-level simulations. In line with the expected trend, in all the examined systems we always have an accuracy improvement when applying more specialized power models. The best performances are provided by model_2, with errors slightly higher than 10% and 5% for at_1_phase and at_4_phase respectively. In any case, all the values reported in Table 3 are in the typical accuracy range of RTL estimation techniques (5%–20%), without evident penalizations.

Table 3. Average estimation errors for different power model specializations from the simulation on SystemC/TLM examples.

Power model	p2p/HS	at_1_phase	at_4_phase
model_0	14.80%	15.90%	12.50%
model_1 (phase)	_	13.80%	8.40%
model_1 (status)	_	12.50%	10%
model_2 (phase+status)	_	10.30%	5.90%

Study on Components Derived From the Soclib Framework

After this study on the examples included in SystemC/TLM releases, we have verified our methodology also on virtual prototypes provided by the SoCLib research project. SoCLib is an open platform for virtual prototyping of multi-processors system on chip described in SystemC language. This platform provides a library of SoC IP components used in concrete applications and modelled at different abstraction levels, such as transaction-level and CABA (cycle accurate – bus accurate). Moreover SoCLib provides several system prototypes built by the aggregation of the library components and executable through the standard SystemC simulation environment. These system prototypes define complex architectures related to real VLSI applications. In the last years SoCLib components have been used as effective benchmarks in several researches to evaluate the performances of digital systems, as in [24] and [25].

In our study we have considered a SoCLib prototype performing fir 128 computations under the control of a processor unit. The corresponding architecture is constituted by four main IP components: a Mips32-based processor, a RAM memory, a computational unit (FIR 128 core) that implements the FIR algorithm and a NoC structure that connects the other components in compliance with the VCI on-chip-bus protocol. We have applied our methodology on this prototype, following the steps illustrated in the study on the SystemC/TLM demonstration examples. First of all, from the IP components provided by SocLib we have derived the descriptions at RTL and gate-level for each of the four components. The realization of such descriptions has required a modeling activity of about three months.
Also in this study we have considered three power model variants: a basic transaction-level power model (model_0), without TLM data specialization, and two more refined power models (model_1 and model_2), specialized with reference to one and two TLM data respectively. In this case, the TLM data considered for power model specialization are transaction status and command type (e.g. read/write operation). In the transaction status we have added also the handling of error conditions, in order to have a better matching with the low-level system representations. In the power simulations on the NoC component, it was not possible to apply model_2, because the tasks of such component are affected only by transaction status, without dependencies on the operations that involve the connected components.

In Table 4 and Table 5 there are reported the different code sizes and the estimation errors for each of the four components constituting the examined architecture. From the results in Table 4, moving from RTL to TL descriptions we have a reduction in modeling effort similar to the one observed in SystemC/TLM examples, with the exception of the NoC component. In this case we have a much more relevant benefit, with a code reduction superior to 60%. This result is a plain consequence of the specific capabilities of SystemC/TLM in the modeling of communication tasks, which represent the primary activity in a NoC structure.

Table 4. Code size of the SoCLib components in the different abstraction levels and TL-RTL effort reduction.

Abstraction level	Processor	RAM	NoC	FIR 128
Gate level	321.2 KB	143 KB	237.5 KB	125.3 KB
RTL	155 KB	94.9 KB	121.4 KB	70.7 KB
TL	101.8 KB	53.5 KB	40 KB	41.5 KB
Effort reduction (RTL−TL)/RTL×100	34%	43%	66%	41,3%

Table 5. Average estimation errors for different power model specializations from the simulations on SoCLib components.

Power model	Processor	RAM	NoC	FIR 128
model_0	30.70%	38.90%	30%	35%
model_1 (status)	26.30%	24%	8.80%	19%
model_1 (command)	10.80%	17.20%	–	13%
model_2 (status+command)	7.70%	10.30%	–	11%

The simulation times of the whole benchmark have confirmed the speedup factors seen in SystemC/TLM examples. The TL system description can be simulated gaining about one order of magnitude in comparison to the RTL counterpart.

As concerns power estimation accuracy, the data in Table 5 show some significant differences with respect to the results achieved from the SystemC/TLM examples. The foremost point is the stronger influence of TLM data in power model performances. As first observation, the estimation error is always excessive and unacceptable when we apply model_0, which does not leverage on TLM data specialization. On the other hand, the simulations with model_1 bring to discordant results. More specifically, this power model provides acceptable errors for the processor, RAM and FIR 128 component when the TLM data specialization concerns the command type. Conversely, the specialization with respect to transaction status leads to rather high errors. In the case of NoC we have an opposite situation, since model_1 provides a good accuracy when the TLM data specialization concerns transaction status.

This mutable behavior can be explained through the different influence of the TLM data in the intrinsic functionalities of the components. To be more precise, the command type is the TLM data with the highest impact on the tasks carried out at run-time by processor, RAM and FIR 128. In comparison with transaction status, command type is more strongly related to the active and power-consuming tasks. For this reason, the command specialization of model_1 leads to better estimations for these components. In the case of NoC we have a different situation, because transaction status is the predominant TLM data and model_1 provides a good estimation error when specialized with respect to transaction status. To be more precise, for this component there is no dependence on command type and transaction status is the unique TLM data affecting the NoC tasks. Accordingly, the best power model specialization can be achieved only in reference to this TLM data.

As expected, for all the components the best results are obtained when applying the most specialized power model (model_2 for processor, RAM and FIR 128; model_1 for NoC). In this case, we always have quite good power estimations with an estimation error around 10%. Such accuracy is at the same level of the best results observed for the SystemC/TLM demonstration examples.

RTL power estimation methodologies provide a variable accuracy, depending on the applied estimation technique. With reference to typical accuracy levels, the average estimation errors may be circumscribed in a percentage range between 5% and 20%. Only advanced cycle-accurate techniques are able to provide errors below 5%. The experimental results show how our methodology, based on a transaction-accurate paradigm, can provide estimation errors in the middle between cumulative and cycle-accurate techniques. However, to address these performances, it is necessary a power model specialization with respect to the TLM data more influent in power dissipation.

Finally, in Table 6 we have reported the overhead due to the application of our methodology in terms of both additional code and extra simulation time. This overhead is evaluated with respect to the basic description in which no power estimation tasks is introduced; the code size of such descriptions has been reported in the TL row of Table 4. Going from left to right, the columns 2–5 show the code added to each component to apply our methodology, i.e. the instructions to define the energy lookup-tables and the run-time querying functions. These instructions have been realized following the implementation guidelines illustrated in Section 4. The bottom values represent the percentage increments with respect to the code of basic descriptions.

Table 6. Overhead of the proposed approach in terms of modeling effort and simulation time.

Power model	Processor	RAM	NoC	FIR 128	Simulation Extra-time
model_0	2.4 KB 2.3%	1.6 KB 3%	1 KB 2.5%	1.8 KB 4.3%	0.20%
model_1 (status)	4 KB 3.9%	2.2 KB 4.1%	1.8 KB 4.5%	2.5 KB 6%	1.50%
model_1 (command)	4.4 KB 4.3%	2.5 KB 4.7%	-	2 KB 4.8%	1%
model_2 (status+command)	11.6 KB 11.4%	5.8 KB 10.8%	-	4.5 KB 10.8%	2.40%

The rightmost column shows instead the extra-time required to simulate the whole system applying the different power model variants on each components. As previously said, the NoC component cannot be run with model1-command and model2 power model. In these cases we have then used model1-status on the NoC component.

The data in Table 6 show how the application of our approach requires a limited modeling effort (below 12% in all the considered cases) and entails a very low penalty in simulation time. These results have been achieved considering properly long simulation times, consisting in the execution of more than 100.000 transactions between the components.
In summary, we can conclude on the validity of the proposed methodology for transaction-level power analysis. Against accuracy analogous to RTL estimation techniques, the higher simulation speed and the lower modeling efforts denote a clear advantage in the design and analysis of complex VLSI systems.

CONCLUSIONS

This paper has presented a power estimation methodology applicable on transaction-level prototypes of VLSI digital systems. The proposed approach is strictly related to the modeling constructs of the SystemC/TLM language, without being restricted to specific communication modalities. These features allow a general and systematic application, with the possibility to refine the estimation accuracy through a power model specialization based on TLM data.

After an introductory overview on the modeling principles of SystemC/TLM, the discussion has dealt with the conceptual aspects of the methodology and the application modalities. Finally, the results coming from an experimental verification have shown the effectiveness of the proposed approach and the design advantages in comparison with RTL estimation techniques. In particular, the power model specialization with respect to TLM data is proved to be a fundamental condition to achieve an acceptable/good estimation accuracy.

REFERENCES

1. L. Cai, D. Gajski, Transaction level modeling: an overviewProceedings of the CODESþISSS Conference 200319–24, in:.
2. C. Piguet, Low-Power CMOS Circuit (Technology, Logic Design and CAD Tools), Taylor & Francis Group, London, 2006, chapter 19.
3. S. Gupta, F. Najm, Analytical models for RTL power estimation of combinational and sequential circuits, IEEE Trans. Comput.- Aided Des. Integr. Circuits Syst. 19 (7) (2000) 808–814.
4. S. Gupta, F. Najm, Power modeling for high-level power estimations, IEEE Trans. VLSI Syst. 8 (1) (2000) 18–29.
5. Bogliolo, R. Cognati, E. Macii, M. Poncino, Parameterized RTL power models for soft macros, IEEE Trans. VLSI Syst. 9 (6) (2001) 880–887.
6. M. Bruno, A. Macii, M. Poncino, RTL power estimation in an HDL-based design flow, IEE Proc. Comput. Digit. Tech. 152 (6) (2005) 723–730.
7. W. Qing, Q. Qinru, M. Pedram, D. Chih-Shun, Cycle-accurate macro-models for RT-level power analysis, IEEE Trans. VLSI Syst. 6 (4) (1998) 520–528.
8. N.R. Potlapally, A. Raghunathan, G. Lakshminarayana, M.S. Hsiao, S.T. Chakradhar, Accurate power macro-modeling techniques for complex RTL circuitsProceedings of the 14th International Conference on VLSI Design 2001235–241, in:.
9. L. Zhong, S. Ravi, A. Raghunathan, N.K. Jha, RTL-aware cycle-accurate functional power estimations, IEEE Trans. Comput.-Aided Des. Integr. Circuits Syst. 25 (10) (2006) 2103–2117.
10. SystemC TLM 2.0 Language Reference Manual. Available Online: ⟨http://www. accellera.org⟩.
11. G. Beltrame, D. Sciuto, C. Silvano, Multi-accuracy power and performance transaction-level modeling, IEEE Trans. Comput.-Aided Des. Integr. Circuits Syst. 26 (10) (2007) 1830–1842.

12. Ahmadinia, B. Ahmad, T. Arslan, Efficient high-level power estimation for multi-standard wireless systems, in: Proceedings of the Symposium on VLSI, Montpellier, 79 April 2008, pp. 275–280.
13. M. Grammatikakis, S. Politis, J.P. Schoellkopf, C. Papadas, System-level power estimation methodology using cycle- and bit-accurate TLM, Design Proceedings of the Automation and Testing in Europe Conference 2011, 2011.
14. N. Dhanvada, L. Ing-Chao, V. Naravanan, A power estimation methodology for SystemC transaction level modelsProceedings of the CODESþISSS Conference 2005142–147.
15. M. Streubur, R. Rosales, R. Hasholzner, C. Haubelt, J. Teich, ESLPower and performance estimation for heterogeneous MPSOCS using SystemC, Forum on Specification & Design Language, FDL, Oldenburg, 1315 September 2011.
16. PKtool software and documentation. Available Online: ⟨http://pktool.source forge.net/⟩.
17. M. Conti, G. Vece, S. Colazilli, Extension of SystemC framework towards power analysis, Forum on Specification & Design Languages, FDL, Sophia Antipolis, 2224 September 2009.
18. G.B. Vece, S. Gigli, S. Orcioni, M. Conti, Power optimization in multi-core system-on-chip, in: George Kornaros (Ed.), Multi-core embedded systems, Taylor & Francis Group LLC CRC Press, 2010, pp. 71–110, chapter 3.
19. G. Vece, M. Conti, Power analysis of embedded systems: the PKtool simulation environment, in: Solutions on Embedded Systems, Springer Netherlands, Lecture Notes in Electrical Engineering, 2011, pp. 301–314 (chapter 21).
20. G. Vece, M. Conti, Power estimation in embedded systems within a SystemCbased design context: the PKtool environment, in: Proceedings of the IEEE Seventh Int. Workshop on Intelligent Solutions in Embedded Systems, WISES09, Ancona, Italy, June 2526, 2009, pp.179184.
21. J. Rabaey, Low Power Design Essential, Springer-Verlag, 2009 chapter 3.
22. H. Lebreton, P. Vivet, Power modeling in SystemC at transaction level, in: Application to a DVFS Architecture, Symposium on VLSI 2008, ISVLSI 2008, Montpellier, 79 April 2008.
23. SoCLib software and documentation. Available Online: ⟨http://www.soclib.fr/ trac/dev⟩.
24. Greiner, S. Taktak, Z. Zhang, A reconfigurable routing algorithm for a faulttolerant 2D-mesh network-on-chip, in: Design Automation Conference, DAC, Anaheim, 813 June 2008.

25. Vieira-de-Mello, I. Maia, A. Greiner, F. Pêcheux, Parallel simulation of SystemC TLM 2.0 compliant MPSoC on SMP workstations, in: Proceedings of the Design, Automation & Test in Europe Conference, DATE, Dresden, 812 March 2010. Giovan

CITATION

G.B. Vece, M. Conti, S. Orcioni, Transaction-level power analysis of VLSI digital systems, Integration, the VLSI Journal, Volume 50, June 2015, Pages 116-126, ISSN 0167-9260, http://dx.doi.org/10.1016/j.vlsi.2015.02.003.

CHAPTER 3

A Buffer Placement Algorithm to Overcome Short-Circuit Power Dissipation in Mesh Based Clock Distribution Network

John Reuben, Mohammed Zackriya V., Harish M. Kittur and Mohd Shoaib

School of Electronics Engineering, VIT University, Vellore 632014, India

ABSTRACT

In the recent past, Mesh-based clock distribution has received interest due to their tolerance to process variations in deep-sub micron technology. Mesh buffers are placed on the mesh to drive the large load capacitance of clock sinks and mesh wire capacitance. In this paper, we propose a buffer placement algorithm which can overcome the short circuit power dissipated in clock meshes. Our buffer placement algorithm uses clustering technique to judiciously place buffers such that short-circuit power is minimized while minimizing skew at the same time. This is verified by Monte carlo simulations incorporating process, voltage and systemic variations in NGSPICE.

INTRODUCTION

Traditionally, tree-based clock distribution has been the preferred method in most VLSI designs. However, in Deep Sub Micron (DSM) technology, clock skew induced by process variations is posing a threat to the reliability of tree-based distribution. Mesh-based distribution, due to their redundant paths offers a variation tolerant alternative to tree-based distribution at the cost of increased power dissipation (due to increased wire capacitance) [1]. Tree driven mesh is a hybrid clock distribution scheme in which a top level clock tree feeds a leaf level mesh. Leaf level clock mesh synthesis is well studied in [1], [2] and [3]. In the leaf level mesh, buffers are placed at the mesh nodes (Fig. 1) to drive the large capacitance (input capacitance of flip flops/registers called clock sinks and mesh wire capacitance).

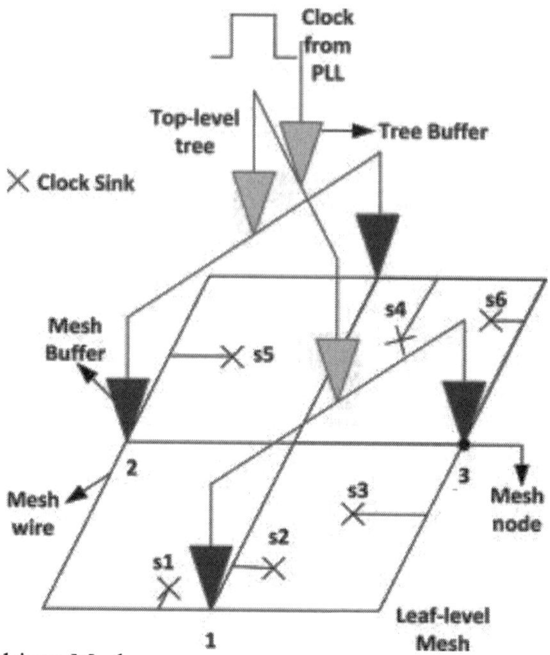

Figure 1. Tree driven Mesh.

EXISTING BUFFER PLACEMENT ALGORITHMS

Buffer placement on the mesh is a crucial step in the design of the clock distribution network. Although buffers can be placed on all the mesh nodes, such a liberal placement of buffers will be an inefficient way (due to power and area consumed by the buffers) to minimize the skew in a mesh. All the previous works ([1], [2] and [3]) place buffers at certain mesh nodes such that the clock latency to all the sinks is equalized. The buffer placed at a particular mesh node is designed to drive the capacitance load around that mesh node.

The buffers are placed using a set-cover algorithm with a discrete buffer library in Ref. [1]. In Ref. [2], the buffers are placed by Iterative Buffer Deletion algorithm (IBD). In Ref. [3], the buffers are placed at the mesh nodes according to the density of sinks in its vicinity and their proximity to the sinks. Since buffers have to be placed in hundreds of locations for designs having thousands of sinks in a clock mesh, the buffer sizing and placement is typically automated. Hence a discrete buffer library, $\mathrm{Library}=\{b_1,b_2,\ldots,b_n\}$ is used. For example, if the load capacitance at mesh nodes varies between 0 and 300 fF in a design, a buffer library of 6 buffers can be used: b_1 to drive 0–50 fF, b_2 to drive 51–100 fF,b_3 to drive 101–150 fF…b_6 to drive 251–300 fF. Each of the buffers is designed/sized to drive its load capacitance with a particular slew constraint. Since clock signals are critical timing signals, they are tightly slewed, typically at 10% of their time period. Hence an ideal buffer to feed a 1 GHz clock to a clock mesh must have a slew no greater than 100 ps when driving its corresponding load.

All the aforementioned algorithms place buffers using a discrete buffer library which may be inefficient i.e to drive a 101 fF load or a 149 fF load, the buffer placement algorithm might place the same buffer-b_3. Buffers contribute to area penalty which is a very costly commodity in sub micron designs. More importantly, there will be Short-Circuit (SC) power dissipation between the buffers in the mesh. As illustrated in Fig. 2, there is a short-circuit path formed between adjacent buffers when the clock reaches the mesh buffers at different times. In this paper, we refer to the power dissipated due to the formation of this short-circuit path as short-circuit power dissipation. This short-circuit power is different from the one inherent in CMOS switching which can be made negligible by careful design. The difference of clock arrival at mesh buffers is attributed to process variations in the top level tree driving the mesh ([1]). The existing buffer placement algorithms overlook this short circuit power dissipation while placing the buffers on the mesh. They only concentrate on achieving minimum skew with minimum number of buffers on the mesh. Hence we need an algorithm which places buffers such that SC power dissipation is minimized while conserving the clock skew. In this paper, we propose a clustering based buffer placement algorithm which has the following properties:

1. Our buffer placement algorithm is short-circuit power 'aware'. Since buffers are placed near the centroids of clusters, they are physically spread apart minimizing the possibility of forming short-circuit paths. This will reduce the SC power dissipation as will be verified by simulations in Section 5.1.
2. Computationally less intensive than the buffer placement algorithms of [1] and [2](Section 5.2)
3. More robust to variations in top-level tree driving the mesh (Section 5.3).

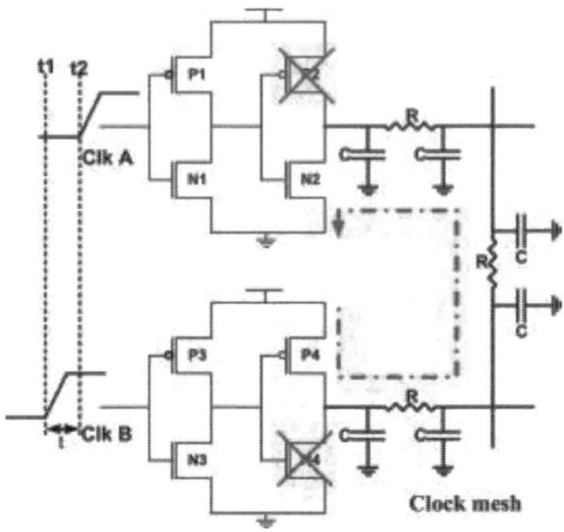

Figure 2. Short circuit path in a clock mesh driven by buffers.

CLUSTERING BASED BUFFER PLACEMENT ALGORITHM

The distribution of clock sinks is highly uneven in real designs. This is evident from the ISPD2010 clock network synthesis benchmarks ([4]) which are based on real 45 nm microprocessor designs of INTEL (Fig. 4). The proposed buffer placement is based on clustering which is a technique to find similarity in data. Clustering is often used in image processing and data mining to find patterns in images/data sets. Here we use clustering to find groups of sinks which can be assigned a single buffer. The detailed steps are:

1. Start with the sink closest to the lower left corner of the mesh. Let it be sink s_1 of the first cluster c_1
2. Find m nearest sinks to s_1 and keep adding them to cluster c_1 till the capacitance of sinks in a cluster reaches a target capacitance, say 100 fF
3. Find the capacitance centroid of the sinks in the cluster c_1 as follows

$$(x_{c1}, y_{c1}) = \left(\frac{\sum_{i=1}^{m} x_i \times C_{si}}{\sum_{i=1}^{m} C_{si}}, \frac{\sum_{i=1}^{m} y_i \times C_{si}}{\sum_{i=1}^{m} C_{si}} \right) \quad (1)$$

where si, $i = 1$ to m represents the m sinks belonging to cluster c_1 and C_{si} represents the capacitance of sink si having co-ordinates (xi, yi).

4. Place a buffer which can drive the target load (100 fF) at the mesh node closest to the capacitance centroid of cluster c_1.
5. Find the next sink closest to the last sink of cluster c_1 and repeat Steps (2)–(4) for the remaining sinks till all sinks are covered by a cluster

Fig. 3 illustrates the clustering algorithm. A direct application of the algorithm will result in clusters spanning large areas when the sinks distribution is very sparse (like cluster c_2). This will lead to increased clock latency which will affect the skew of the mesh. To avoid this, an imaginary box is used internally in Step (2) to limit the cluster area even when the target capacitance is not reached.

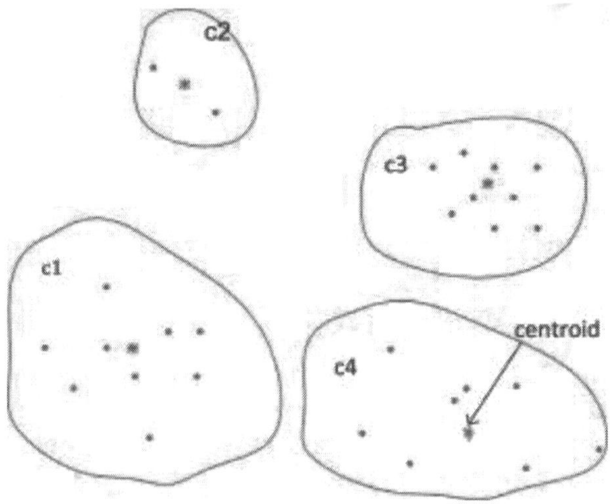

Figure 3. Formation of clusters of clock sinks which can be assigned a single buffer.

MESH FORMATION AND BUFFER PLACEMENT FOR BENCHMARK CIRCUIT

In our previous work ([3]), we presented a capacitance driven mesh formation algorithm which forms a minimum wire-length, non-uniform mesh according to the density of capacitance in the chip. Clock sinks are connected to the mesh edges by a combination of steiner tree and stubs. Benchmarks 6,7 and 8 of ISPD2010 clock network synthesis contest have 981, 1915 and 1134 clock sinks respectively. The clock sinks in these benchmarks are unevenly distributed making buffer placement challenging. We formed the capacitance driven mesh of [3] for these benchmarks. We then applied our proposed buffer placement algorithm and placed the buffers at mesh nodes. Fig. 4 shows benchmark6 with capacitance driven mesh formed and buffers placed according to our proposed clustering algorithm (target capacitance of 100 fF and 200 fF). The buffers are placed only at those mesh nodes near the capacitance centroids of clusters.

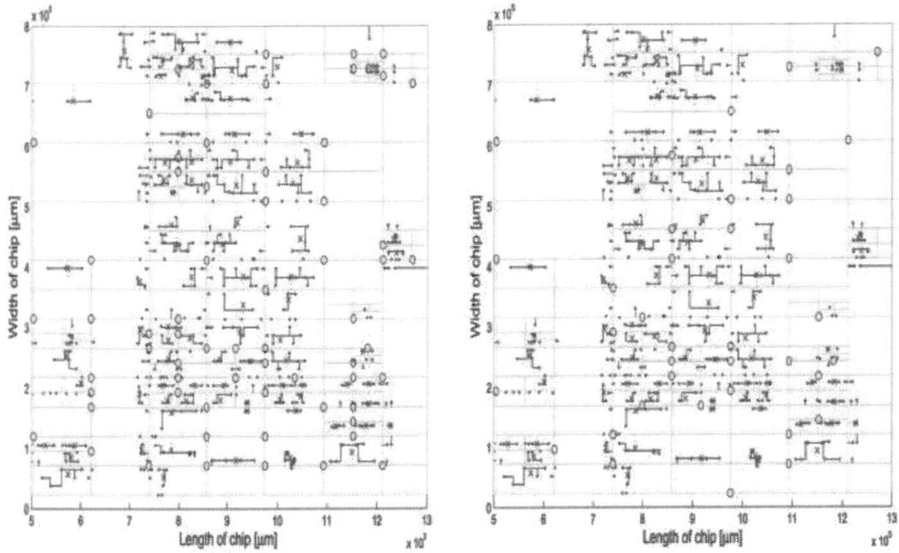

(a) Buffer placement: 100fF/cluster (b) Buffer placement: 200fF/cluster

Figure 4. Portion of ISPD2010 benchmark:'.' denotes clock sinks, 'o' denotes buffers.

MESH MODELING AND SIMULATION

The mesh wire has got resistance and capacitance. We use a distributed model of the wire which is an accurate model when compared to the lumped modeling. According to [5], the lumped RC model is a pessimistic model for a resistive-capacitive wire. So we used the π-model to model the mesh wires. Because of the importance of the distributed model, most circuit simulators have a built-in distributed rc models of high accuracy. NGSPICE supports the uniform-distributed rc-line model (URC) of Berkeley's SPICE3 simulator. This model approximates the rc line as a network of lumped RC segments with internally generated nodes. The mesh model is shown in Fig. 5.

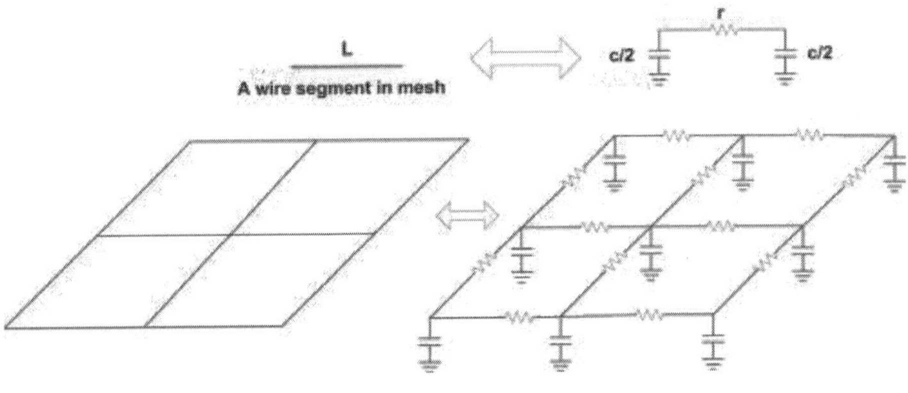

Figure 5. Mesh model.

Clock mesh simulation is computationally intensive due to the thousands of wire segments which have to π-modeled. Sliding window scheme ([6]) reduces the computational complexity by splitting the mesh into windows where the mesh is accurately modeled. Outside the window, the mesh is approximately modeled. i.e the resistance and capacitance of wires joining the sinks to the mesh wires (stubs) are ignored. The clock latency is evaluated for the sinks inside the window and then the window is moved to evaluate the latency at all the other sinks ([3]). We have split the entire mesh into 4 windows as illustrated in the simulation flow (Fig. 7).

Simulation Conditions and Results

The simulation conditions are tabulated in Table 1. 45 nm PTM files were used to simulate the mesh buffers. The clock signal feeding the mesh has a slew of 50 ps to model the degradation in the clock reaching the mesh buffers. We studied benchmark6 for 4 different cases of buffer placement – target capacitance of clusters being 100 fF, 150 fF, 200 fF and 250 fF. For each case, we placed an appropriately sized buffer which can drive the load without slew violation (Table 2).

Table 1. Simulation conditions and variations.

Parameter	Value
V_{DD}	1 V
Clock frequency	1 GHz
Resistance of mesh wire	0.3 Ohm/μm
Capacitance of mesh wire	0.16 fF/μm
Parameter	Variation (Standard deviation $\sigma = 5\%$ of nominal)
Channel length	Gaussian with 3σ variation
V_{th}	Gaussian with 3σ variation
V_{DD}	Gaussian with 3σ variation
Inter-buffer skew	50 ps

Table 2. Sizes of buffers to drive different cluster sizes.

Case	Capacitance/cluster	First inverter			Second inverter		
		W/L of P1 [nm]	W/L of N1 [nm]	*Wp*/*Wn*	W/L of P2 [nm]	W/L of N2 [nm]	*Wp*/*Wn*
1	100 fF	82/45	57/45	1.438	1656/45	1177/45	1.407
2	150 fF	114/45	82/45	1.390	2451/45	1718/45	1.426
3	200 fF	151/45	110/45	1.372	3258/45	2281/45	1.428
4	250 fF	189/45	130/45	1.454	4073/45	2852/45	1.428

To study the effectiveness of our buffer placement and to accurately predict the skew post-fabrication, we performed Monte Carlo simulations with statistical variations. Threshold voltage V_{th} and channel length L_{eff} are the two key device parameters that are subject to variation in DSM technology [7]. We did not consider variations in interconnects since transistor variations are dominant compared to interconnect variations [8]. Hence we varied V_{th} and channel length L_{eff} by 5% of their nominal value according to a gaussian distribution. V_{DD} variation is also incorporated. As already stated, the clock signal will not reach all the mesh buffers at the same time due to process variations in the top level tree driving the mesh. This will lead to short circuit power dissipation between adjacent buffers in the mesh (Fig. 2). To incorporate this phenomenon in simulation, the clock arrival time at the mesh buffers is varied randomly such that the maximum difference in arrival times between any 2 buffers does not exceed 50 ps. This is called 'Inter-buffer skew' and illustrated in Fig. 6 (the 'delay' parameter of the PULSE command in NGSPICE is varied to implement different arrival times of the clock at different buffers).

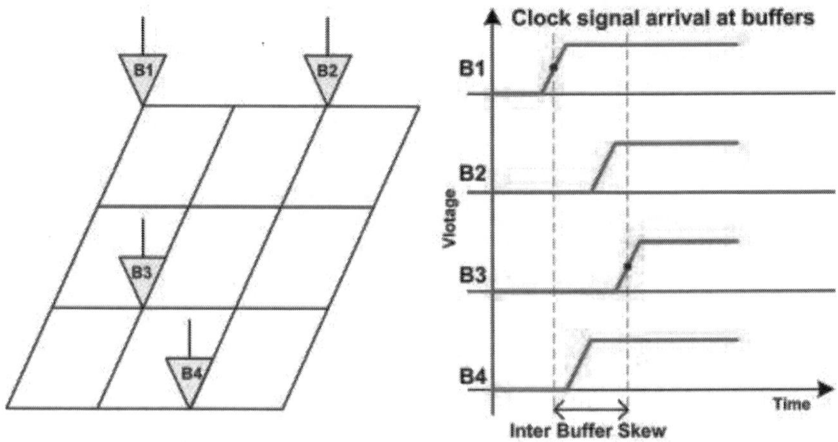

Figure 6. Inter-buffer skew: The maximum difference in clock arrival time at buffers.

Following the simulation flow in Fig. 7, we performed 200 Monte carlo (MC) simulations on benchmark 6, 7 and 8. The global skew was calculated from the clock latency at sinks. The total power was calculated by integrating the current drawn from V_{DD}. Table 3 shows the average skew and average power dissipation for the 4 different cases of buffer placement on benchmark6 after 200 MC simulations. We observe that bigger the capacitance/cluster, lesser buffers have to be placed. But the skew increases due to increased imbalance in the path between buffers and sinks. Interestingly, the total power dissipation remains the same whether 116 buffers or 60 buffers are placed on the mesh. This is because the total load (C_{load}) to be driven by all the buffers together remains the same (total load is the capacitance of sinks and the capacitance of the mesh wires) for a particular benchmark. Hence the dynamic power dissipation ($C_{load} \cdot V_{DD}^2 \cdot f$) will remain the same. As depicted in Table 3, when the capacitance/cluster is 100 fF, 116 buffers are needed while if the capacitance/cluster is 200 fF, only 60 buffers are needed because each buffer in this case is of a larger size compared to the 100 fF case (Table 2). Since the total power dissipation remains constant irrespective of the number of buffers (Table 3), it implies that short circuit power component of total power does not change with the increase in the number of buffers.

This proves that the short circuit power component in Table 3 is due to the 50 ps inter-buffer skew and not due to the increased number of buffers i.e whether 60 buffers or 116 buffers are placed, our buffer placement doesn't aggravate short-circuit power consumption.

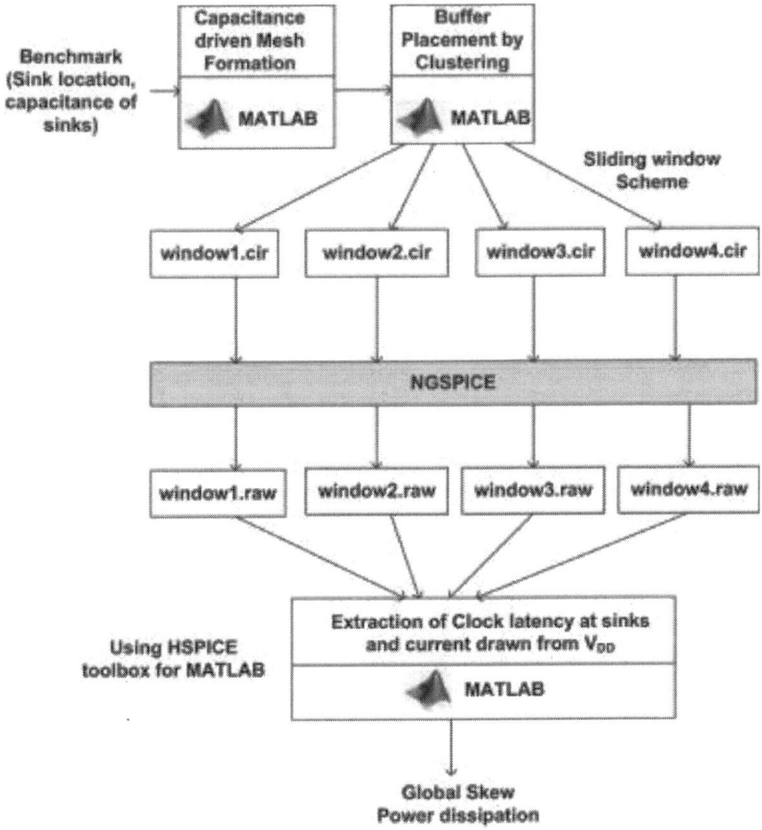

Figure 7. Simulation flow.

Table 3. Skew and Power averaged over 200 MC simulations for benchmark6.

Capacitance/cluster	No. of buffers	Clock skew	Total power dissipation
100 fF	116	26.35 ps	21.77 mW
150 fF	79	29.18 ps	21.64 mW
200 fF	60	34.9 ps	21.88 mW
250 fF	47	38.95 ps	21.41 mW

The skew and power dissipation for benchmarks 6,7,8 for clustering based buffer placement (for capacitance/cluster of 100 fF) under the simulation/variation conditions ofTable 1 is tabulated in Table 4. Among the buffer placement algorithms published in literature, the Iterative Buffer Deletion (IBD) of [2] attains the best skew and power dissipation on ISPD2010 benchmarks. As observed in Table 4, our buffer placement achieves lower power dissipation compared to IBD of [2] for benchmark6 and almost the same power dissipation for benchmarks 7 and 8, at the cost of increased skew. Although our buffer placement is inferior to [2] in terms of skew, our algorithm is computationally much less complex than [2].

Table 4. Performance of buffer placement of [2] vs Clustering based buffer placement.

	IBD of [2]		Ours	
Benchmark	Skew (ps)	Power (mW)	Skew (ps)	Power (mW)
Benchmark6	12	26.09	26.35	21.77
Benchmark7	7	44.71	27.76	44.1
Benchmark8	7	31.96	26.58	31.9

Computational Complexity of Buffer Placement Algorithms: A Qualitative Comparison

[1] uses a set-cover algorithm to place the buffers. In this algorithm, the cost of placing a buffer at a mesh node is calculated using a cost function and buffers are placed such that all the sinks are covered with minimum cost. The time complexity of the algorithm is $O(n \cdot \beta)$ where n is the number of candidate mesh nodes and β is the size of the buffer library ([9]). Guthaus [2] uses an iterative buffer deletion algorithm for buffer placement. In this algorithm, minimum sized buffers are placed at all the mesh nodes initially. Then, buffers are sized using sensitivity-based linear programming method. Smaller sized buffers are then removed iteratively till the skew exceeds a certain user specified bound. According to Ref. [2], the time complexity of their algorithm is 7 times more than that of the set-covering approach of [1]. In our approach, the time complexity is highly reduced because we dont do buffer sizing i.e we use uniformly sized buffers throughout the mesh. The capacitance driven mesh forms a non-uniform mesh according to the density of clock sinks. Then, the clustering based buffer placement algorithm forms clusters of sinks according to a particular target capacitance and places uniformly sized buffers on the

mesh. Hence, in general, the time complexity is O(k), where k is the number of sinks. Practically, the run-time complexity of our algorithm is the time it takes to form the clusters and find the capacitance centroid of each cluster. For benchmark7 having 1915 sinks, this time was around 7.6 s when our buffer placement algorithm was run on a 3.3 GHz Intel i3 processor with 2 GB RAM. Hence our algorithm has a much lower time complexity compared to the buffer placement of [1] and [2]. We are not able to compare our time complexity with the buffer placement of [2] quantitatively since CPU runtime of buffer placement alone is not provided in Ref. [2].

Performance of the Clock Mesh with Increased Inter-Buffer Skew

To probe the tolerance of our mesh to inter-buffer skew, we performed some simulation studies with increased inter-buffer skew. According to Ref. [10], the SC power dissipation in a clock mesh is a linear function of inter-buffer skew. Wilke [11] conducted a study to find the % contribution of this SC power to the total power as a function of inter-buffer skew (The clock arrival time at the mesh buffers was varied randomly between 0 and inter-buffer skew). The study concluded that the SC power dissipation can be as high as 50% of the total power when the inter-buffer skew is 15% of the clock period (150 ps for a 1 GHz clock). To study our clock mesh design in light of [11] and [12]'s observations, we increased the inter-buffer skew from 50 ps to 150 ps in steps of 25 ps for benchmark6 and performed 200 MC simulations as earlier. As the inter-buffer skew at the mesh buffers increases, the skew and power in the mesh increase only marginally as shown in Fig. 8. In fact, even for 100 ps inter-buffer skew which implies the worst case top-level tree (the acceptable skew in a clock tree must be $\leq$10% of clock time period) for 1 GHz clock, the total power dissipation increases by only 1.42 mW when compared to the 50 ps case. This proves the tolerance of our design to inter-buffer skew. For 150 ps inter-buffer skew, the total power dissipation increases by 3.82 mW when compared to the 50 ps case. This increase is justified because of the unreasonable input skew at mesh buffers. Such unreasonable input skew will happen in the case a very poor top-level tree design which can be avoided. Hence our buffer placement algorithm is tolerant to inter-buffer skew of acceptable measure. We performed the same study on benchmark7 and plotted our observations in Fig. 9. Due to the large size of benchmark7, we performed only 100 MC simulations and varied inter-buffer skew in steps of 50 ps. Irrespective of the benchmark, our buffer placement is tolerant to inter-buffer skew and power consumption doesn't increase rapidly with increase in inter-buffer skew, as predicted by [11].

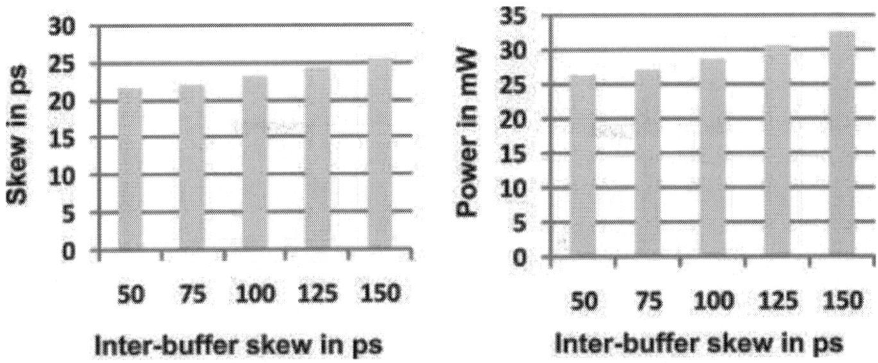

Figure 8. Skew and power averaged over 200 MC simulations of benchmark6 with different Inte buffer skew.

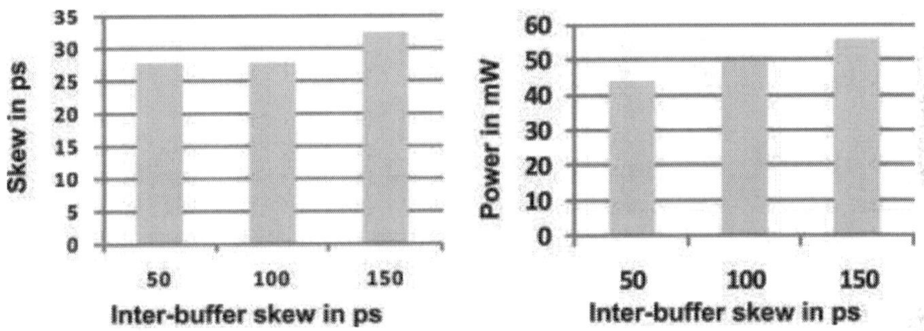

Figure 9. Skew and power averaged over 100 MC simulations of benchmark7 with different Inter-buffer skew.

Finally, we compared the tolerance of our buffer placement with the tolerance of buffer placement of [3] by performing 200 MC simulations for an inter-buffer skew of 150 ps. While the buffer placement of [3] consumed 35.11 mW for 50 ps inter-buffer skew on benchmark6 ([3]), the total power dissipation grew rapidly to 50.96 mW for 150 ps inter-buffer skew i.e on the same mesh and under the same simulation conditions, the total power dissipation of [3] increases by 45.14% while that of the proposed work increases by 17.54% only. All the above observations establish the robustness of the proposed buffer placement algorithm.

CONCLUSION

Power dissipation in a clock mesh is increased due to short circuit power dissipation in mesh buffers. By forming clusters of sinks and placing buffers at the mesh nodes near the centroid of the clusters, we are able to minimize the short-circuit power dissipation in the mesh. When buffers are placed near centroids of clusters, they are physically spread apart, thus avoiding formation of short-circuit paths when the clock arrival time varies at the mesh buffer inputs. This comes at the cost of some increased skew but the power dissipation is reduced since short-circuit component of total power is greatly minimized. Although our buffer placement algorithm achieves the same power dissipation as the recently proposed algorithm ([2]), it is better than [2] in terms of computational complexity. We also verified that with the increase in inter-buffer skew (from 50 ps to 150 ps), the skew and power dissipation of our mesh doesn't increase rapidly but only marginally. Hence clustering based buffer placement is also robust against variations in the top-level tree driving the mesh.

REFERENCES

1. Rajaram, D. Pan, Meshworks: a comprehensive framework for optimized clock mesh network synthesis, IEEE Trans. Comput. Aided Design Integr. Circuits Syst. 29 (12) (2010) 1945e1958.
2. M.R. Guthaus, X. Hu, G. Wilke, G. Flach, R. Reis, High-performance clock mesh optimization, ACM Trans. Des. Autom. Electron. Syst. 17 (3) (2012) 1e33, 33.
3. J. Reuben, V.M. Zackriya., S. Nashit, H.M. Kittur, Capacitance driven clock mesh synthesis to minimize skew and power dissipation, IEICE Electron. Express 10 (24) (2013).
4. C.N. Sze, ISPD 2010 high performance clock network synthesis contest: benchmark suite and results, in: Proceedings of the 19th International Symposium on Physical Design, ISPD '10, ACM, 2010, p. 143.
5. Jan M. Rabaey, Digital Integrated Circuits, a Design Perspective, Prentice Hall, 2003.
6. H. Chen, C. Yeh, G. Wilke, S. Reddy, H. Nguyen, W. Walker, R. Murgai, A sliding window scheme for accurate clock mesh analysis, in: Computer-aided Design, 2005. ICCAD-2005. IEEE/ACM International Conference on, 2005, pp. 939e946.
7. S. Sarangi, B. Greskamp, R. Teodorescu, J. Nakano, A. Tiwari, J. Torrellas, Varius: a model of process variation and resulting timing errors for microarchitects, IEEE Trans. Semicond. Manuf. 21 (1) (2008) 3e13.
8. S. Abe, M. Hashimoto, T. Onoye, Clock skew evaluation considering manufacturing variability in mesh-style clock distribution, IEICE Trans. Fundam. Electron. Commun. Comput. Sci. E91-A (12) (2008) 3481e3487.
9. G. Venkataraman, Zhuo Feng, Jiang Hu, Peng Li, Combinatorial algorithms for fast clock mesh optimization, IEEE Trans. Very Large Scale Integr. (VLSI) Syst. 18 (1) (2010) 131e141.

10. Vaisband, E.G. Friedman, R. Ginosar, A. Kolodny, Low power clock network design, J. Low Power Electron. Appl. 1 (1) (2011) 219e246.
11. G.R. Wilke (Ph.D. thesis), Analysis and Optimization of Mesh-based Clock Distribution Architectures, vol. 100e101, Universidade Federal do Rio Grande do Sul., 2008.
12. Gustavo Wilke, Renan Fonseca, Cecilia Mezzomo, Ricardo Reis, A novel scheme to reduce short-circuit power in mesh-based clock architectures, in: Proceedings of the 21st Annual Symposium on Integrated Circuits and System Design (SBCCI '08), 2008, pp. 117e122.

CITATION

John Reuben, Mohammed Zackriya V., Harish M. Kittur, Mohd Shoaib, A buffer placement algorithm to overcome short-circuit power dissipation in mesh based clock distribution network, Engineering Science and Technology, an International Journal, Volume 18, Issue 2, June 2015, Pages 135-140, ISSN 2215-0986, http://dx.doi.org/10.1016/j.jestch.2014.10.005.

CHAPTER 4

Power Conditioning Using Dynamic Voltage Restorers under Different Voltage Sag Types

Ahmed M. Saeed[1], *Shady H.E. Abdel Aleem* [2], *Ahmed M. Ibrahim*[1], *Murat E. Balci* [3], *Essam E.A. El-Zahab*[1]

[1] Electrical Power and Machines Engineering, Cairo University, Giza 12613, Egypt
[2] Mathematical, Physical and Life Sciences, 15th of May Higher Institute of Engineering, 15th of May City, Cairo, Egypt
[3] Electrical and Electronics Engineering, Balikesir University, Balikesir, Turkey

ABSTRACT

Voltage sags can be symmetrical or unsymmetrical depending on the causes of the sag. At the present time, one of the most common procedures for mitigating voltage sags is by the use of dynamic voltage restorers (DVRs). By definition, a DVR is a controlled voltage source inserted between the network and a sensitive load through a booster transformer injecting voltage into the network in order to correct any disturbance affecting a sensitive load voltage. In this paper, modelling of DVR for voltage correction using MatLab software is presented. The performance of the device under different voltage sag types is described, where the voltage sag types are introduced using the different types of short-circuit faults included in the environment of the MatLab/Simulink package. The robustness of the proposed device is evaluated using the common voltage sag indices, while taking into account voltage and current unbalance percentages, where maintaining the total harmonic distortion percentage of the load voltage within a specified range is desired. Finally, several simulation results are shown in order to highlight that the DVR is capable of effective correction of the voltage sag while minimizing the grid voltage unbalance and distortion, regardless of the fault type.

INTRODUCTION

Recently, much attention has been focused on the power quality domain in power system networks. Power quality measures the fitness of electric power transmitted from the utilities to the different consumers in the case of the conventional centralized generation or in some cases from the consumers to the utilities in the case of distributed generation. Voltage distortion that may occur due to power system harmonics and voltage sags is widely recognized as the most severe issue affecting power quality, because it affects both the utility company and consumers alike. Nonlinear loads create voltage and current harmonics which may have detrimental effects on consumers' equipment [1], [2] and [3].

IEEE Standard 1159-1995 defines voltage sags as a root-mean-square (rms) variation with a magnitude between 10% and 90% of nominal voltage and duration typically ranging from a few milliseconds to sixty seconds [4]. Voltage sag takes place at nearby feeders with a detrimental feeder subjected to one of the causes of voltage sag. Short circuits due to faults in the power system structure, lightning strokes, high starting currents of induction motors, and inrush currents are the common causes of voltage sags [5]. Voltage sags can be symmetrical or unsymmetrical depending on the causes of the sag. If the individual phase voltages are equal and the phase relationship is 120 degrees, the sag is symmetrical. Otherwise, the sag is unsymmetrical. A three-phase short-circuit fault can produce symmetrical sags. Single line-to-ground, phase-to-phase, or two phase-to-ground faults due to lightning, animals, accidents, and other causes, as well as energizing of large transformers, can cause unsymmetrical sags [6].

A power conditioner is a device proposed to enhance the quality of the power that is delivered to a sensitive electrical load. In addition, it can be defined as a device that acts on delivery of a voltage of the appropriate level and characteristics to facilitate the effective utilization of critical loads. At the present time, one of the power conditioners most commonly used to mitigate voltage sags is the dynamic voltage restorer (DVR). By definition, a DVR is a controlled voltage source inserted between the network and a sensitive load through a transformer injecting voltage into the network in order to correct any disturbance affecting the sensitive load voltage [2] and [5]. More functions can be included with the DVR such as reactive power compensation, harmonics mitigation, and fault current limitations.

DVRs' controllers have an important effect on the system dynamic response, stability and steady-state accuracy [7], [8], [9], [10], [11], [12], [13] and [14]. In the literature, there are many types of controllers that can be used in the DVR compensation practice, such as feedback and feed-forward [8], double-vector [9], proportional and integral (PI) [10], fuzzy and adaptive PI-fuzzy controllers [11] and [12], which are widely used in low-voltage small capacity DVR applications. Recently, a novel software phase-locked loop (SPLL) is proposed by combining the advantages of least-error-squares (LES) filters and the instantaneous symmetrical components method, which has a fast phase-lock tracking ability and guarantees no data fluctuation of the sag detection algorithm under non-sinusoidal conditions [13]. Additionally, a new strategy with the positive and the negative sequence extractions (PNSE) from the fundamental and the higher distorted harmonic orders is proposed [14], which improves the dynamic response of the DVR with an accurate steady-state compensation. Despite the valuable development added by such novel algorithms, they are mainly dedicated to high/medium-voltage applications which need large capacity dynamic voltage restorers with enhanced capability controllers, especially for the utilities that have complex non-typical industrial consumers and may considerably suffer from parameters uncertainty and/or wide range of operation circumstances, such as the grids integrated with large-scale wind and/or solar power resources.

In this paper, modelling of a DVR using PI controller for voltage correction using MatLab software is presented. The pre-sag compensation method has been used as the control strategy to maintain the voltage at the terminals of a sensitive load at its rated value. In other words, the voltage injected by the DVR will be the difference between the voltage at the point of common coupling before and during the sag [2] and [5].

The performance of the device under different voltage sag types is described, where the voltage sag types are introduced using the different types of short-circuit faults included in the environment of the MatLab/Simulink package. The robustness of the proposed device is evaluated using the common voltage sag indices described in [15] and the voltage and current unbalance percentages given in [16], where maintenance of the total harmonic distortion percentage of the load voltage in a specified range complying with IEEE Standard 519-1992 is desired [17]. Finally, several simulation results are shown in order to highlight the viability of the proposed device.

THE PROPOSED METHODOLOGY

Dynamic Voltage Restorer

A dynamic voltage restorer is a solid-state power electronic switching device which is connected in series to the load voltage bus in order to inject a dynamically controlled voltage. This voltage can remove any detrimental effects of a bus fault on a sensitive load voltage.

Fig. 1 shows a schematic diagram of a typical DVR structure which is used for voltage recovery. It consists of the following units:

i. Energy storage unit: This is DC storage energy with a proper capacity which supplies the DVR during compensation by the required real power. It can be simply a capacitor or a battery. Recently, super capacitors, fly wheels, and super-magnet conductors are emerging as energy storage devices with a fast response. Unfortunately, difficult maintenance and the high cost of these facilities compared with conventional facilities have been noted in the power quality markets, delaying their widespread deployment in a broad sense.

ii. Injection transformer: The DVR transfers the voltage which is required for the compensation from the voltage source converter to the distribution network through the injection transformer [18]. The high voltage side is normally connected in series with the distribution network while its low voltage side is connected to the power circuit of the DVR.

iii. Voltage source converter (VSC): This is a power electronic configuration which is used to generate a sinusoidal voltage with the required magnitude, phase, and frequency. Its dc input is the energy storage unit.

iv. LC passive filter: A simple output filter composed of passive elements such as a resistance *R*, inductance *L*, and a capacitance *C*. It is used to reduce the undesired harmonic components of the waveform generated by the converter to their permissible limit. Its output is a sinusoidal waveform with low total harmonic distortion.

v. Bypass switch: This is used to isolate the DVR from the system in case of high currents [19].

vi. Control unit: This is used to detect the presence of voltage sags in the system. In other words, it is considered as a monitor of the load-bus voltage. If a sag voltage is sensed, the controller will be initiated in order to inject the missing voltage after determination of its magnitude and phase [20], [21] and [22].

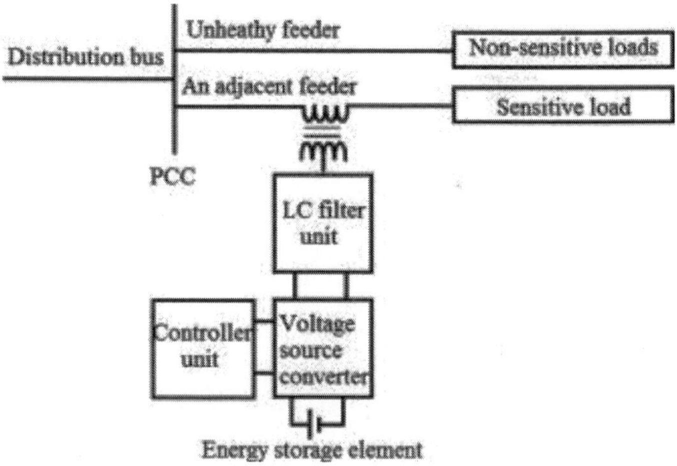

Figure 1. Schematic diagram of a typical DVR structure.

The DVR has two main modes of operation, which are as follows:

Standby mode: This is the monitoring action of the load-bus voltage. No voltage is injected and the transformer low-voltage side is shorted through the converter.

Injection mode: The DVR in this mode injects the required voltage to the system to correct the sag [20].

Consequentially, one can say that the DVR is a series-connected device between the source and a sensitive load that injects a dynamically controlled voltage and protects voltage-sensitive equipment from sags. On the other hand, uninterruptible power supplies (UPSs), static voltage compensators (SVCs), distributed static compensators (DSTATCOMs), and super-magnetic energy storage (SMES) are other approaches that can handle the case. Unfortunately, large size, difficult maintenance, and the higher cost of these facilities compared with the DVR facility have been noted in the markets. Thus, the simplest and cheapest device for voltage sag correction is the DVR.

Voltage Sag Calculation

Fig. 2a shows a simplified equivalent circuit of a Thevenin source system represented by voltage source V_S and source reactance X_S. It is feeding two equal loads represented byZ_1 and Z_2 through two feeders F_1 and F_2, where Z represents the load impedance and $X_F$the magnitude of feeder reactance. I_S is the supply current. In normal operation, the pre-sag voltage at the point of common coupling ($V_{\text{Pre-sag}}$) and the supply current are given as follows:

$$V_{\text{Pre-sag}} = V_S - I_S X_S \tag{1}$$

$$I_S = I_1 + I_2 = \frac{V_{\text{Pre-sag}}}{Z_1 + X_{F1}} + \frac{V_{\text{Pre-sag}}}{Z_2 + X_{F2}} \tag{2}$$

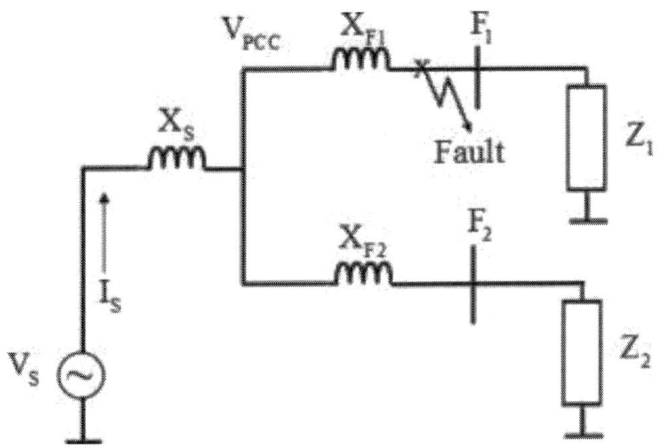

Figure 2a. Simplified equivalent circuit for voltage sag calculation.

When a fault occurs on F_1 (the unhealthy feeder), a high current will flow through it as well as the supply current. During such a case, the supply current $I_{S,\text{fault}}$ and the voltage at the point of common coupling during sag (V_{Sag}) will be given as follows:

$$V_{\text{Sag}} = V_S - I_{S,\text{fault}} X_S \tag{3}$$

$$I_{S,\text{fault}} = \frac{V_{\text{Sag}}}{X_{F1}} + \frac{V_{\text{Sag}}}{Z_2 + X_{F2}} \tag{4}$$

Accordingly, the voltage across the adjacent feeder F_2 will be reduced due to the excessive voltage drop that will appear across the source reactance X_S. This voltage drop will be defined as voltage sag [5]. Hence, a DVR represented by a controlled voltage source V_{DVR} will be inserted between the point of common coupling and the sensitive load Z_2, as shown in Fig. 2b.

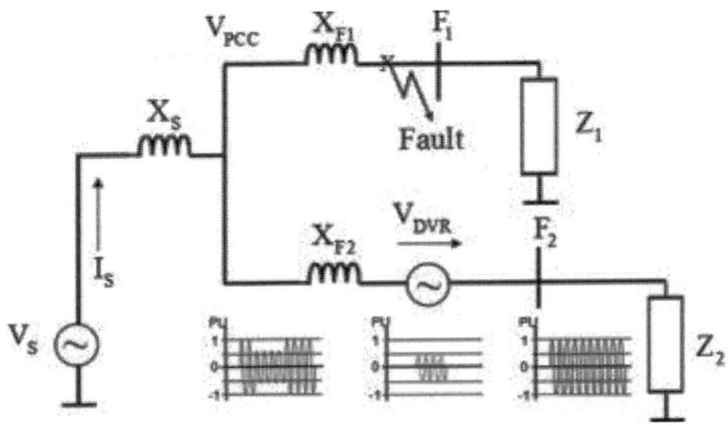

Figure 2b. Simplified equivalent circuit for the DVR voltage injection.

Description of the System under Study

Fig. 2c shows a single line diagram of the system configuration under study. It is composed of a 13 kV, 50 Hz generation system feeding two transmission lines through a three-winding transformer connected in $Y_g/\Delta/\Delta$, 13/66/66 kV. Such transmission lines feed two distribution networks through two transformers connected in Δ/Y_g, 66/0.38 kV. Bus-A represents the unhealthy feeder in which different faults will occur at point *X*, while bus-B represents the adjacent feeder connected to sensitive loads. To validate the performance of the DVR for voltage correction, balancing, and harmonics mitigation, different fault types will be applied at point *X* for the duration of 135 ms. The DVR is simulated to be in operation only for the duration of the fault [5].

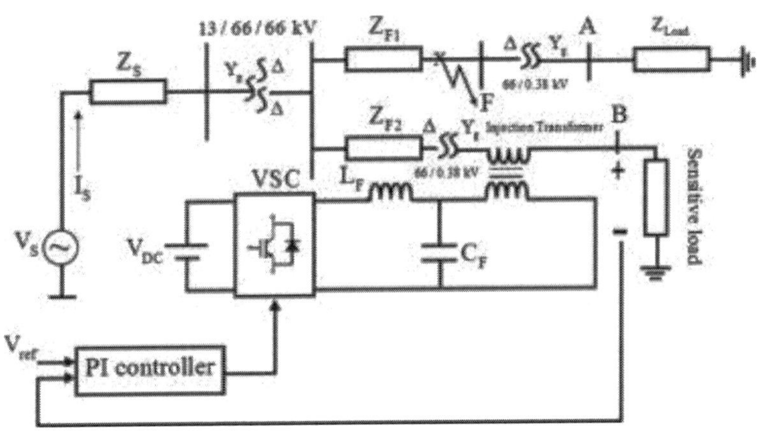

Figure 2c. Single line diagram of the system under study.

The following indices regarding the goodness of the system quality after compensation are taken into consideration:

Voltage Total Harmonic Distortion (THDV)

Harmonic distortion is a good indication of the quality of the system output voltage. According to IEEE 519-1992, THDV for a voltage level up to 69 kV is less than or equal to 5.0% of the fundamental. The maximum individual frequency voltage harmonic is limited to 3% for a system without a major parallel resonance at one of the injected harmonic frequencies [17]. The expression of the THDV measured at the point of common coupling can be written as follows, where *h* is the harmonic number presented and *p* is the phase order so that *p* = *a*, *b*, *c* [23] and [24]:

$$\mathrm{THDV} = \frac{\mathrm{THDV}_a + \mathrm{THDV}_b + \mathrm{THDV}_b}{3} \tag{5}$$

$$\mathrm{THDV}_p = \frac{\sqrt{\sum_{h\geqslant 2} V_{ph}^2}}{V_{p1}} \cdot 100 \tag{6}$$

Voltage sag Indices

Sag indices are indicators that describe the quality of voltage sag or voltage recovery. These indices are sensitive for any disturbance; hence they can give accurate feedback on the system performance. The following indices are discussed in the simulated system results:

Detroit Edison Sag Score (SS)

This is the first voltage sag index that is used in contracts between utilities and consumers. The Detroit Edison Sag Score (SS) is defined as follows [15]:

$$\mathrm{SS} = 1 - \frac{V_a + V_b + V_b}{3} \tag{7}$$

where V_a, V_b, and V_c are the root-mean-square (rms) values of the phase voltages per unit. An *SS* value closer to 0 indicates a good recovered voltage after compensation.

Voltage Sag Lost Energy Index (VSLEI)

During voltage sag, the voltage is below normal for some period of time, which reduces the energy delivered to the loads. This index gives the lost energy W during a sag event, which is defined as follows:

$$W = \sum_{p=a,b,c} W_p = \sum_{p=a,b,c} T_p * \left(1 - \frac{V_p}{V_{\text{nominal}}}\right)^{3.14} \tag{8}$$

where V_p is a phase voltage per unit with respect to the nominal voltage V_{nominal} during the sag event, and T_p is the sag duration in milliseconds for each phase [15] and [16]. It is obvious that for three-phase calculation, the lost energy is added for all three phases.

Phase unbalance rate

The IEEE definition [25] of voltage unbalance, known as the phase voltage unbalance rate (PVUR), is given by

$$\text{PVUR} = \frac{\text{Maximum}(V_d)}{V_{\text{avg}}} \cdot 100 \tag{9}$$

where V_d represents phase voltage deviation from the average phase voltage (V_{avg}).

DVR Control Algorithm

The control algorithm produces a three-phase reference voltage to the series converter that seeks to keep the load voltage at its reference value. In this paper, the well knowndq_0 transformation (Park's transformation) is used to control the DVR as shown in (10), where ω is the angular frequency in radians per second. Hence, the three-phase system is simplified and can be easily controlled after transformation of the three phase *abc*voltage into the two voltage components V_d and V_q. For simplicity, zero phase sequence components are ignored. The control block diagram with a phase-locked loop (PLL) is illustrated in Fig. 2d. Basically, the PLL circuit is used to generate a unit sinusoidal wave in phase with the main voltage [20].

$$\begin{bmatrix} V_d \\ V_q \\ V_0 \end{bmatrix} = \frac{1}{3} \begin{bmatrix} 2\sin\omega t & 2\sin\left(\omega t - \frac{2\pi}{3}\right) & 2\sin\left(\omega t + \frac{2\pi}{3}\right) \\ 2\cos\omega t & 2\cos\left(\omega t - \frac{2\pi}{3}\right) & 2\cos\left(\omega t + \frac{2\pi}{3}\right) \\ 1 & 1 & 1 \end{bmatrix} \begin{bmatrix} V_a \\ V_b \\ V_c \end{bmatrix} \tag{10}$$

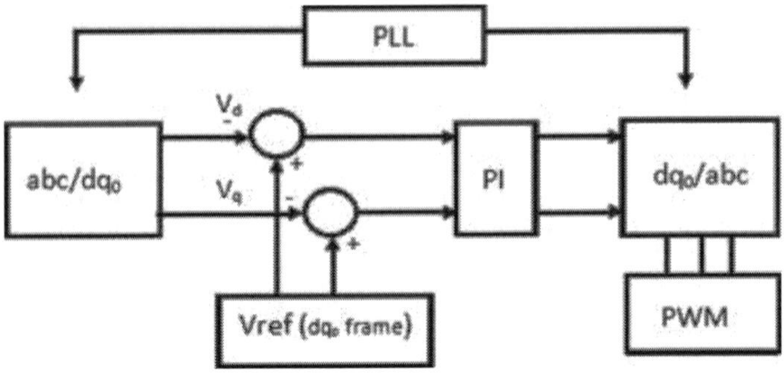

Figure 2d. Control scheme of the dynamic voltage restorer.

In order to maintain simplicity, the control algorithm can be summarized as follows [10]:

- The DVR controller monitors the load-bus voltage. Consequentially, this voltage is transformed to its corresponding *dq* components. The components of the load voltage are compared with the reference voltage *dq* components.
- If a sag voltage is sensed, an error signal will be generated due to the difference between the measured and reference voltage values and the controller will be initiated in order to inject the missing voltage. This error signal drives a PI (proportional and integral) controller which controls the system depending on the actuating error signal. It should be noted that the output signal generated from the PI controller is transformed back to three phase *abc* voltage before it is forwarded to the Pulse Width Modulation (PWM) generator, as shown in Fig. 2d.
- For getting smooth and sinusoidal output voltage, a simple output filter composed of passive elements is needed at the output of the converter. The input to the filter is high frequency modulated 50 Hz AC input. The switching signal that modulates the 50 Hz signal is taken to be 5.5 kHz for the proposed case. Accordingly, a low pass LC filter that suppresses most of the generated harmonic frequencies, is proposed. This will result in a nearly sinusoidal output voltage. Sizing of the filter parameters is given in Table 1[27].

Table 1. System and DVR parameters.

Parameter	Value
Line resistance (Ω)	1
Line inductance (mH)	5
Line frequency (Hz)	50
Load phase voltage (V)	220
Load power per phase (W)	100
Load inductive reactive power per phase (kvar)	0.2
Load capacitive reactive power per phase (kvar)	0.5
Injection transformer turns ratio	1:01
Saw-tooth carrier wave frequency (*Hz*)	5500
DC supply voltage (V)	200
Filter series inductance (mH)	80
Filter series resistance (Ω)	0.1
Filter shunt capacitance (μF)	6
Filter shunt resistance (Ω)	60

The Pulse Width Modulation (PWM) control technique is applied for inverter switching so as to generate a three-phase 50 cycles-per-second sinusoidal voltage at the terminals of the load. Consequentially, its output signal controls the pulses for the inverter. In other words, the PWM generator will generate pulses to trigger the PWM inverter with the desired firing sequence. IGBT is the switching device that is used with the VSC for the DVR operation because a freewheeling diode is connected in antiparallel with each IGBT, thus bringing more flexibility to the proposed device with a compromise among conducting and switching losses [26].

SIMULATION RESULTS AND DISCUSSION

Three cases of an industrial plant (Table 1) were simulated using MatLab software Simulink. The numerical data were taken from an example in an existing publication [20]. The cases under study can be summarized as follows:

- **Case I**: A three-phase to ground fault is applied. Therefore, the system is under type A sag voltage. The voltage during this event is equal in the three phases. This type of sag is considered a balanced sag regardless of the transformer or the load connection.

- **Case II**: A double line to ground fault is applied. Basically, types E, F, and G are only expected if the fault is a double line to ground fault. However, the transformer between the faulted point and the considered bus is a D/Y_g transformer. This means that the system is under type F sag voltage.
- **Case III**: A single line to ground fault is applied. Basically, types B, C, and D are only expected if the fault is a single line to ground fault. The healthy phases show the same magnitude and direction in their voltage change. In this case, the system is under type C sag voltage.

Each fault is applied on the unhealthy feeder for a duration of 0.05–0.185 s (135 ms), respectively. Thus a sag voltage will be sensed on the sensitive feeder.

Table 2 shows the uncompensated system results to be defined and compared with the DVR compensation results.

Table 2. Uncompensated system results in the three cases.

Parameters and cases	Case I	Case II	Case III
$THDV_a$ (%)	4.11	3.16	0.93
$THDV_b$ (%)	8.08	8.13	0.96
$THDV_c$ (%)	8.56	2.84	0.41
THDV (%)	6.92	4.71	0.77
PVUR (%)	8.1	16.3	1.46
SS (%)	50	24.2	1.85
VSLEI (J)	47.45	17.2	0.01

Table 3 shows the compensated system results after using the proposed device. It is obvious that the THDV percentages are consistent with the total harmonic distortion limit of the IEEE Standard 519-1992, for all cases under study. Besides, it is notable, as shown in Table 3, that the phase voltage unbalance rate percentages have met the IEEE recommendations defined in [25].

Table 3. Compensated system results in the three cases.

Parameters and cases	Case I	Case II	Case III
$THDV_a$ (%)	0.46	0.98	0.55
$THDV_b$ (%)	1.75	1.6	0.54
$THDV_c$ (%)	1.7	1.1	0.01
THDV (%)	1.3	1.23	0.37
PVUR (%)	0.5	0.69	0.02
SS (%)	2	0.7	0.22
VSLEI (J)	0.002	0.0003	2E-06

Even when the sag score was high as shown in Case I (50%) or moderate as in Case II (24.2%), or only a slight action for voltage correction and regulation was needed as in Case III (1.85%), the proposed device was able to achieve the required goals while minimizing voltage harmonic distortion, sag lost energy, and voltage unbalance rate.

Seeking a clear demonstration of the proposed device, Fig. 3a, Fig. 3b, Fig. 3c, Fig. 4a,Fig. 4b, Fig. 4c, Fig. 5a, Fig. 5b and Fig. 5c show waveforms of the faulty feeder voltage, the uncompensated sensitive feeder voltage, and the compensated load voltage for all cases under study, respectively. The significant sag voltage recovery in the load voltage waveforms after compensation is obvious.

Case I: Three-phase to ground fault.

Case II: Double line to ground fault.

Case III: Single line to ground fault.

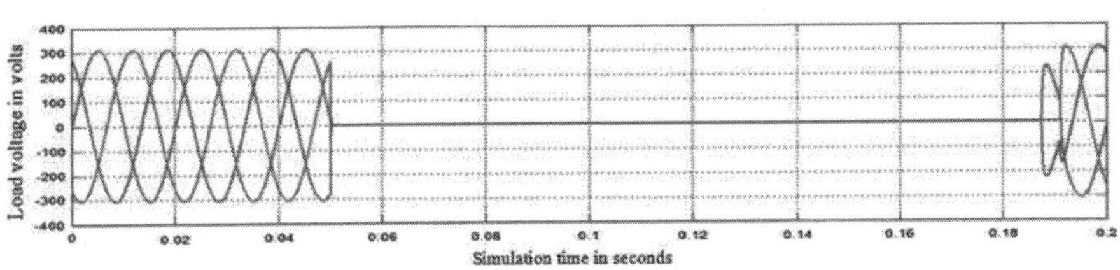

Figure 3a. Faulty feeder voltage in volts versus time in seconds, Case I.

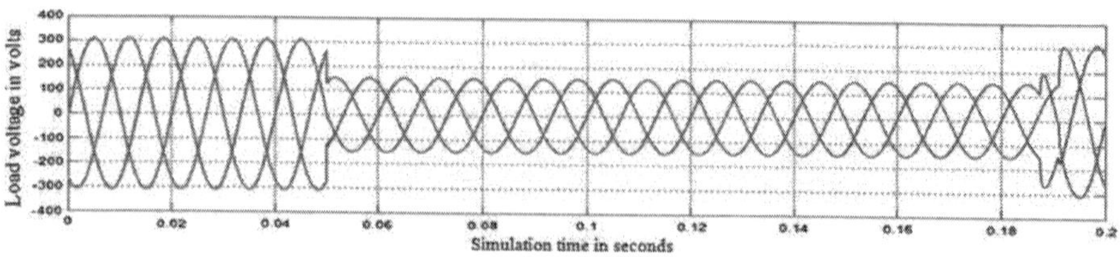

Figure 3b. Uncompensated sensitive feeder voltage in volts versus time in seconds, Case I.

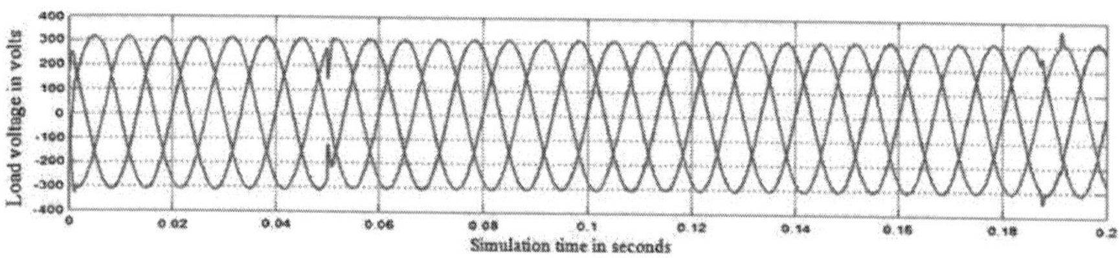

Figure 3c. Compensated load voltage in volts versus time in seconds, Case I.

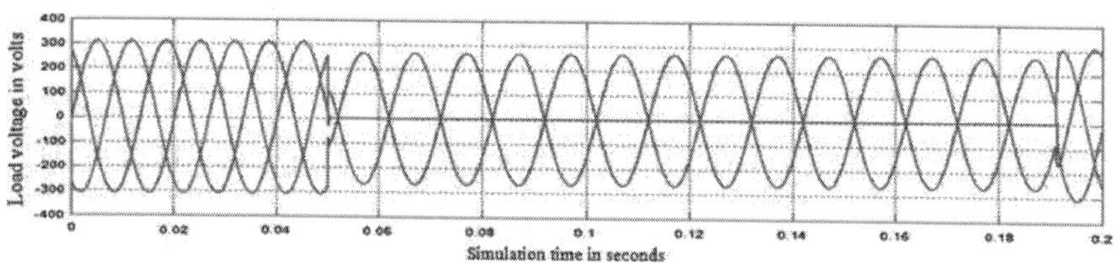

Figure 4a. Faulty feeder voltage in volts versus time in seconds, Case II.

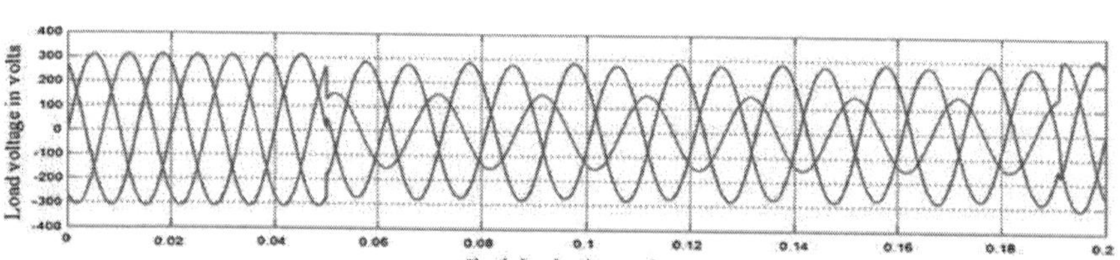

Figure 4b. Uncompensated sensitive feeder voltage in volts versus time in seconds, Case II.

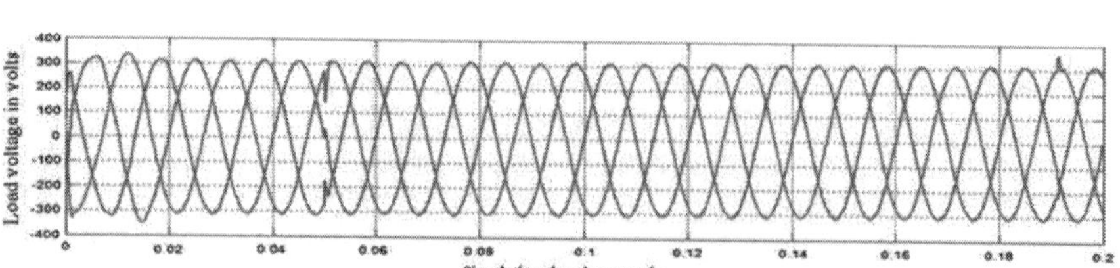

Figure 4c. Compensated load voltage in volts versus time in seconds, Case II.

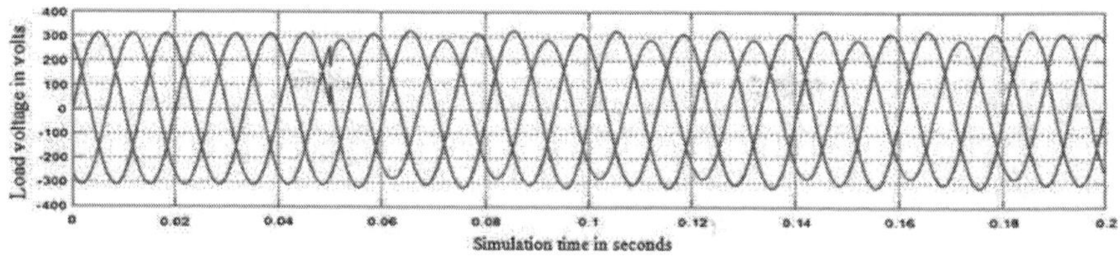

Figure 5a. Faulty feeder voltage in volts versus time in seconds, Case III.

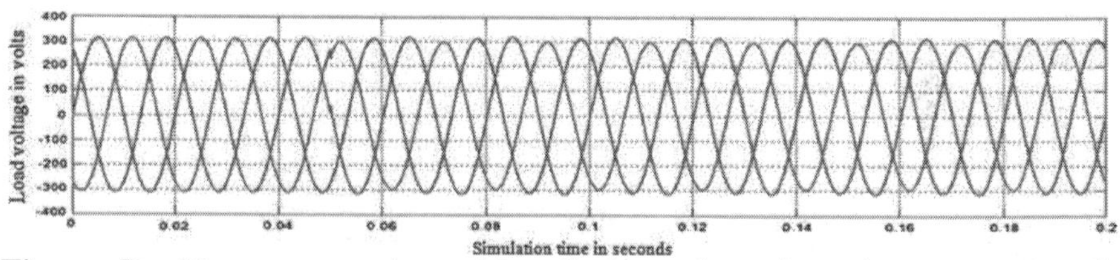

Figure 5b. Uncompensated sensitive feeder voltage in volts versus time in seconds, Case III.

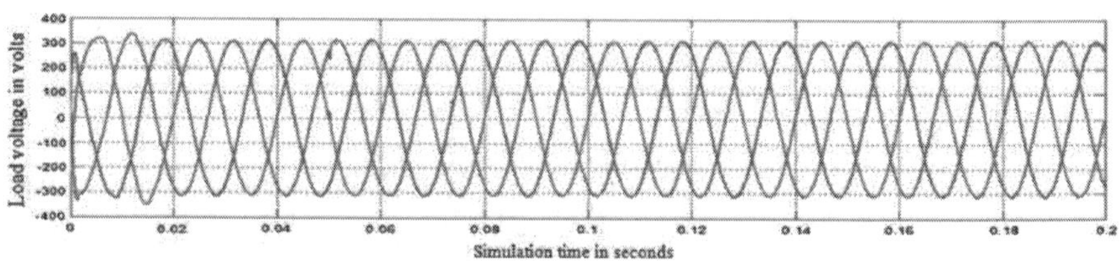

Figure 5c. Compensated load voltage in volts versus time in seconds, Case III.

Fig. 6a shows the voltage total harmonic distortion percentages generated after compensation with respect to changing the energy storage capacity. It is obvious that 200 V_{DC} is the optimal value that achieves the lowest voltage harmonic distortion percentage. Fig. 6b shows the variation of the sag score percentage (%SS) with respect to changing the energy storage capacity. Once more, it is obvious that 200 V_{DC} was an appropriate choice which achieves a low sag score percentage. Additionally, Fig. 6cshows variation of the phase voltage unbalance rate (%PVUR) with respect to changing the energy storage capacity. It is clearly obvious that 130 V_{DC} is the lowest voltage capable of complying with IEEE Standard 112-1991 recommendations with a phase unbalance rate less than 2%.

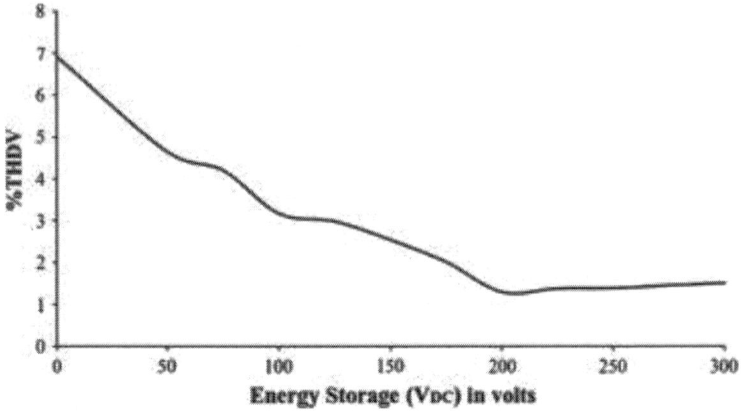

Figure 6a. %THDV variation versus the energy storage unit.

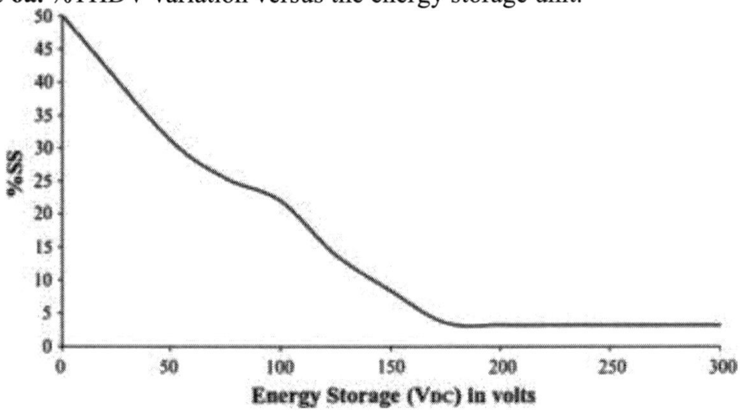

Figure 6b. %SS variation versus the energy storage unit.

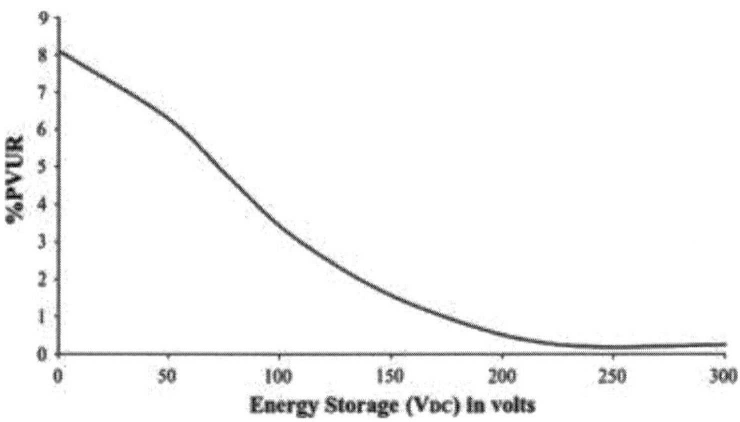

Figure 6c. %PVUR variation versus the energy storage unit.

In order to check the control robustness against harmonic distortion, simulation results of the system with an additional three-phase diode rectifier type (a typical non-linear load) which is connected in parallel with the sensitive load, are shown in Table 4 and Table 5for the uncompensated and compensated systems, respectively. Even with the more harmonic-distorted system due to the connection of the non-linear load, it is notable that the THDV percentages for the compensated system are consistent with the total harmonic distortion limit of the IEEE Standard 519-1992, for all cases under study. Besides, it is notable, as shown in Table 5, that the phase voltage unbalance rate percentages have met the IEEE recommendations (less than 2%). Thus, the proposed device with the presented control scheme was able to achieve the required goals of voltage recovery while minimizing voltage harmonic distortion, sag lost energy, and voltage unbalance rate. This is mainly because the DVR has similar structure and principle of operation to series active filters [26]; accordingly, it operates well in harmonics mitigation. Finally, many controllers can be used for the DVR. Each controller has its own advantages and disadvantages. For example, choosing appropriate controller for the DVR is greatly affected by the percentage of load distortion; if it is small, the PI controllers may be the most appropriate scenario because of their simplicity; if it is moderate; the fuzzy or adaptive PI controllers can be a good choice; if it is high, then the controllers previously reported [13] and [14] are the best because of their effectiveness in operating under non-sinusoidal conditions. Accordingly, more simulation results should be done to determine which controller can considerably enhance the DVR's dynamic response with a reasonable efficiency and without decreasing accuracy, while taking into account the various power quality factors. The findings of this research will be presented in a future paper.

Table 4. Uncompensated system results with linear and non-linear loads in the three cases.

Parameters and cases	Case I	Case II	Case III
$THDV_a$ (%)	5.47	3.95	4.4
$THDV_b$ (%)	7.47	9.66	5.64
$THDV_c$ (%)	9.59	3.34	5.55
THDV (%)	7.51	5.65	5.2
PVUR (%)	7.89	14.56	1.23
SS (%)	50.61	30.98	12
VSLEI (J)	47.72	18.55	0.55

Table 5. Compensated system results with linear and non-linear loads in the three cases.

Parameters and cases	Case I	Case II	Case III
$THDV_a$ (%)	0.81	2.4	1.18
$THDV_b$ (%)	1.84	3.03	1.08
$THDV_c$ (%)	1.97	1.29	1.65
THDV (%)	1.54	2.24	1.3
PVUR (%)	0.52	1.36	0.12
SS (%)	2.05	2.21	0.31
VSLEI (J)	0.002	0.007	0

CONCLUSIONS

Voltage sag detection is the first step in enabling a proper solution to many disturbances affecting a power system network. In this paper, modelling of DVR for voltage correction using MatLab software is presented. The performance of the device under different voltage sag types is described. Various power quality indices are used to evaluate the performance of the grid with the proposed device. Several simulation results are introduced to validate that the proposed DVR operation scheme fulfils the required goals. It is obvious that the DVR is capable of effective correction of the voltage sag while minimizing the grid voltage unbalance and harmonics distortion, regardless of the fault type.

ACKNOWLEDGEMENTS

The authors gratefully acknowledge and thank the team of the Electrical Power and Machines Engineering Department, Faculty of Engineering, Cairo University for their helpful comments and support.

REFERENCES

1. Zobaa AF, Abdel Aleem SHE. A new approach for harmonic distortion minimization in power systems supplying nonlinear loads. IEEE Trans Ind Inform 2014;10(2):1401–12.
2. Pakharia A, Gupta M. Dynamic voltage restorer for compensation of voltage sag and swell: a literature review. Int J Adv Eng Technol 2012;4(1):347–55.

3. Rozlan MBM, Zobaa AF, Abdel Aleem SHE. The optimisation of stand-alone hybrid renewable energy systems using HOMER. Int Rev Electr Eng 2011;6(4B):1802–10.
4. IEEE Standard 1159-1995. IEEE recommended practice for monitoring electric power quality; 1995.
5. Wahab SW, Yusof AM. Voltage sag and mitigation using dynamic voltage restorer (DVR) system. Elektrika 2006;8(2): 32–7.
6. Guasch L, Corcoles F, Pedra J. Effects of symmetrical and unsymmetrical voltage sags on induction machines. IEEE Trans Power Deliv 2004;19(2):774–82.
7. Abdul Rahman S, Janakiraman PA, Somasundaram P. Voltage sag and swell mitigation based on modulated carrier PWM. Int J Electr Power Energy Syst 2015;66:78–85.
8. Cheng PT, Chen JM. Design of a state-feedback controller for series voltage-sag compensators. IEEE Trans Ind Appl 2009;45(1):260–7.
9. Awad H, Blaabjerg F. Transient performance improvement of static series compensator by double vector control. In: 19th Annu IEEE Applied Power Electron Con and Expo, Anaheim, CA, USA, IEEE; 2004. p. 607–13.
10. Hazarika S, Roy SS, Baishya R, Dey S. Application of dynamic voltage restorer in electrical distribution system for voltage sag compensation. Int J Eng Sci 2013;2(7):30–8.
11. Babu PS, Kamaraj N. Performance investigation of dynamic voltage restorer using PI and fuzzy controller. Int Conf on Power, Energy and Control, Sri, Rangalatchum, Dindigul, IEEE; 2013. p. 467–72.
12. Ezhilarasan S, Balasubramanian G. Dynamic voltage restorer for voltage sag mitigation using PI with fuzzy logic controller. Int J Eng Res Appl 2013;3(1):1090–5.
13. Chen G, Zhang L, Wang R, Zhang L, Cai X. A novel SPLL and voltage sag detection based on LES filters and improved instantaneous symmetrical components method. IEEE Trans Power Electron 2015;30(3):1177–88.
14. Chen G, Zhu M, Cai X. Medium-voltage level dynamic voltage restorer compensation strategy by positive and negative sequence extractions in multiple reference frames. IET Power Electron 2014;7(7):1747–58.
15. Bollen MHJ. Voltage sag indices, Draft 1.2. A working document for IEEE P1564 and CIGRE WG 36-07. <http://grouper.ieee.org/ groups/sag/ documents.html>.
16. IEEE Standard 1459-2010. Definitions for the measurement of electric power quantities under sinusoidal, non-sinusoidal, balanced or unbalanced conditions; 2010.
17. IEEE Standard 519-1992. IEEE recommended practices and requirements for harmonic control in electrical power systems; 1992.
18. Wang B, Venkataramanan G, Illindala M. Voltage restorer using transformer coupled H-bridge converters. IEEE Trans Power Electron 2006;21(4):1053–61. [19] Marefatjou H, Sarvi M. Compensation of single-phase and three-phase voltage sag and swell using dynamic voltage restorer. Int J Appl Power Eng 2012;1(3):129–44.

19. Saeed AM, Abdel Aleem SHE, Ibrahim AM, El-Zahab EEA. Power quality improvement and sag voltage correction by dynamic voltage restorer. Int Rev Autom Control 2014;7(4):386–93.
20. Choi SS, Li BH, Vilathgamuwa DM. Dynamic voltage restoration with minimum energy injection. IEEE Trans Power Syst 2000;15(1):51–7.
21. Damor AK, Babaria VB. Voltage sag control using DVR. Nat Conf Recent Trends in Engineering Technology, Gujarat, India; 2011. p. 1–4.
22. Abdel Aleem SHE, Zobaa AF, Sung ACM. On the economical design of multiple-arm passive harmonic filters. In: 47th International Universities' Power Engineering Conf, UPEC'12, Uxbridge, Middlesex, UK, IEEE; 2012. p. 1–6.
23. Balci ME, Zobaa AF, Abdel Aleem SHE, Sakr S. An algorithm for optimal sizing of the capacitor banks under non-sinusoidal and unbalanced conditions. Recent Pat Electr Eng 2014;7(2):116–22.
24. IEEE Standard 112-1991. IEEE standard test procedure for polyphase induction motors and generators; 1991.
25. Akagi H. New trends in active filters for power conditioning. IEEE Trans Ind Appl 1996;32(6):1312–22.
26. Abdel Aleem SHE, El-Mathana MT, Zobaa AF. Different design approaches of shunt passive harmonic filters based on IEEE std. 519–1992 and IEEE std. 18–2002. Recent Pat Elec Eng 2013;6(1):68–75.

CITATION

Ahmed M. Saeed, Shady H.E. Abdel Aleem, Ahmed M. Ibrahim, Murat E. Balci, Essam E.A. El-Zahab, Power conditioning using dynamic voltage restorers under different voltage sag types, Journal of Advanced Research, Available online 6 March 2015, ISSN 2090-1232, http://dx.doi.org/10.1016/j.jare.2015.03.001

CHAPTER 5

MOS-FET as a Current Sensor in Power Electronics Converters

Rok Pajer [1], Miro Milanovič [1], Branko Premzel [2] and Miran Rodič [1,]

[1]Faculty of Electrical engineering and computer sciences, University of Maribor, Smetanova 17, SI-2000 Maribor, Slovenia
[2]Piktronik, Cesta k Tamu 17, SI-2000 Maribor, Slovenia

ABSTRACT

This paper presents a current sensing principle appropriate for use in power electronics' converters. This current measurement principle has been developed for metal oxide semiconductor field effect transistor (MOS-FET) and is based on UDS voltage measurement. In practice, shunt resistors and Hall effect sensors are usually used for these purposes, but the presented principle has many advantages. There is no need for additional circuit elements within high current paths, causing parasitic inductances and increased production complexity. The temperature dependence of MOS-FETs conductive resistance RDS−ON is considered in order to achieve the appropriate measurement accuracy. The "MOS-FET sensor" is also accompanied by a signal acquisition electronics circuit with an appropriate frequency bandwidth. The obtained analogue signal is therefore interposed to an A-D converter for further data acquisition. In order to achieve sufficient accuracy, a temperature compensation and appropriate approximation is used ($R_{DS-ON}=R_{DS-O}N(\vartheta_j)$). The MOS-FET sensor is calibrated according to a reference sensor based on the Hall-effect principle. The program algorithm is executed on 32-bit ARM M4 MCU, STM32F407.

INTRODUCTION

Recently, there has been significant interest in any kind of power electronics' converters for using current sensors for current control purposes [1,2,3]. The more frequently-used current measurement principles in power electronics' converter applications are based on the magnetic principle and/or voltage measurement on a shunt resistor. Both of them must also be equipped with appropriate analogue signal acquisition electronic circuits. However, high currents, flowing through shunt resistors, causes additional power losses. On the other hand, the shunt measurement principle reduces the costs of power electronic converters, whilst those sensors based on the magnetic principle are relatively expensive. Based on the shunt resistor principle, MOS-FETs (basic elements in power converters) can be used as current sensors. MOS-FETs are appropriate as sensors in many different application, as was investigatedin [4,5,6]. The on-state resistance of MOS-FETs, R_{DS-ON}, can act as the shunt resistance and can be used as it. The benefit of such organized measurement devices is obvious: there is no need for additional elements for current measurement purposes in power converters.

Motivations for this research effort have included cost savings associated with reducing the number of current sensors. Some of the basic challenges associated with achieving effective and accurate current measurement are dealing with temperature compensation, linearization and the calibrating of a whole measurement chain [7,8]. The use of MOS-FETs as a current sensor is considered in [9], where the authors describe an analog solution to be used for transistor protection. Furthermore, in [10], the authors applied the MOS-FET as a sensor, placing attention on the temperature circuit, as well, but an additional MOS-FET was used as a sensor circuit. The synchronization issues must also be considered when using the micro-controller. Such an organized sensor has limited accuracy, depending on the quality of the sensor's instrumental amplifier and the evaluation of the R_{DS-ON} changes. The sensor circuit has limited bandwidth and disturbance rejection. RDS−ONdepends on gate-source voltage UGS and MOS-FET junction temperature. The range of resistance changes due to the temperature is significant, which means that the MOS-FET sensor's accuracy strongly depends on junction temperature at a particular time instant. The MOS-FET junction temperature (ϑj) is impossible to measure; therefore, the thermal model is used for evaluation of the $R_{DS-ON} = R_{DS-ON}(\vartheta_j)$. This phenomena was studied in [11,12].

This paper proposes an algorithm for using the MOS-FET resistance RDS−ON for current measurement purposes in a power converter circuit. The Introduction is followed by Section 2, where the measuring principle is presented considering the estimation of the MOS-FET junction temperature by evaluating of the MOS-FET power dissipation. Furthermore, an appropriate acquisition electronics circuit is featured for UDS, case and heat sink temperature measurement. Section 3 describes the experimental setup. Experimental results and the discussion are given in Section 4. Section 5 concludes the paper.

THE MEASURING PRINCIPLE

Voltage UDS is measured when the MOS-FET is in the on-state. Figure 1a shows the half-leg circuit with the appropriate inductor as a DC-DC step-up converter. Voltage over Q_2 terminals (drain source) appears as a consequence of the current flowing through transistor Q_2 on-state resistance R_{DS-ON}, as indicated in Figure 1b. Figure 2 shows the expected waveforms of inductor current Id, MOS-FET current IDS and the voltage on the MOS-FET U_{DS}. The upper MOS-FET Q_1 is in the off-state and the lower Q_2 in the on-state. When digital control is applied, the current sensor is usually triggered in the middle of the time interval Ton. Due to measurement synchronization, the measured signal corresponds to the average value of the inductor current at the interval Ts indicated by Id,av. If the resistance R_{DS-ON} is known by measuring the U_{DS} in the middle of interval Ton, by applying Ohm's law, the average current can be calculated as follows:

$$I_{d,av} = \frac{U_{DS}}{R_{DS-ON}} \qquad (1)$$

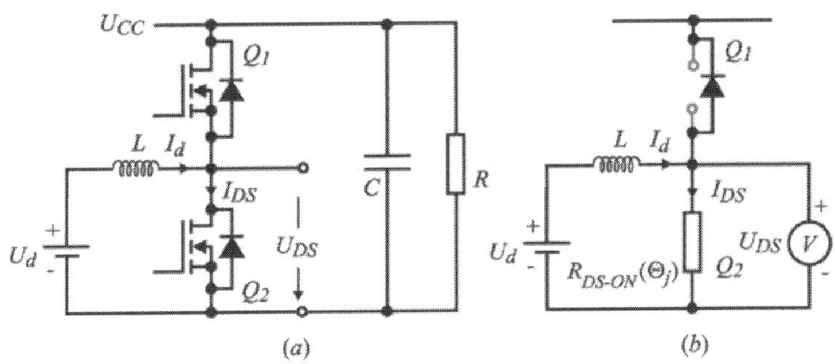

Figure 1. The step-up converter scheme.

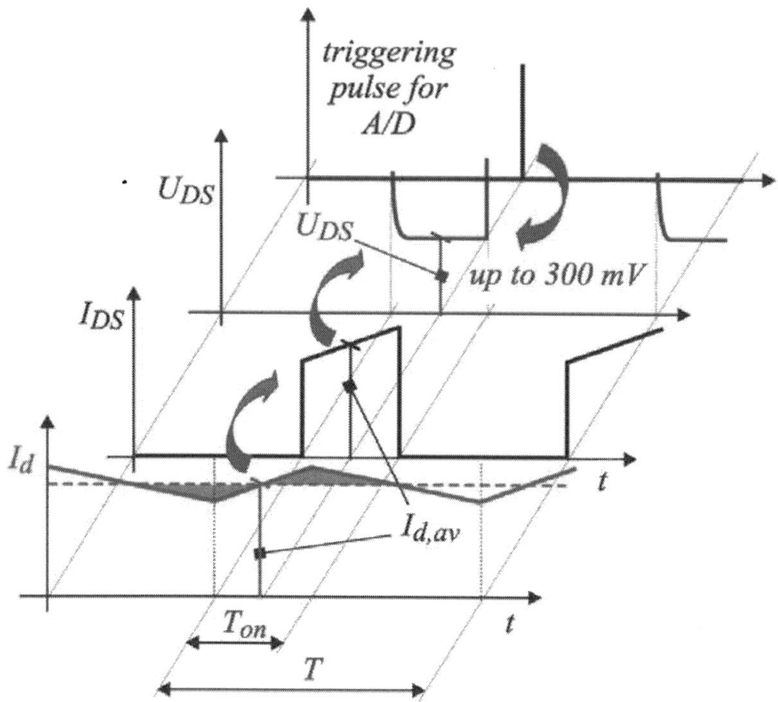

Figure 2. Synchronization of the UDSmeasurement.

In general, the RDS−ON is known, but it is not constant. In order to determine the RDS−ON accurately, it is necessary to evaluate its dependence on the junction temperature ϑj. The data-sheet [13] of the chosen MOS-FET transistor IRFB4110 was used for the evaluation of the function $R_{DS-ON} = R_{DS-ON}(\vartheta_j)$The temperature dependence was evaluated by the curve shown in Figure 3. Those points indicated by * were extracted from the curve and are arranged in Table 1, where *r*represents the normalized MOS-FET resistance. For the chosen MOS-FET:

$$r = R_{DS-ON}(\vartheta_j)/R_{DS-ON}(25^oC) \tag{2}$$

where RDS−ON(25oC)=3.7mΩ. As suggested in [14] and by using the data indicated in Table 1, the approximated analytical term for *r* is obtained by a second order quadratic equation with coefficients K0, K1 and K2 as follows:

$$r = K_0\vartheta_j^2 + K_1\vartheta_j + K_2 \tag{3}$$

where K0, K1 and K2 are indicated in Table 2. As follows from the above analyses, the crucial issue for appropriate estimation of on-state resistance RDS–ON is evaluation of the junction temperature ϑj, because the value of RDS–ON can change by 300% according to the ϑj.

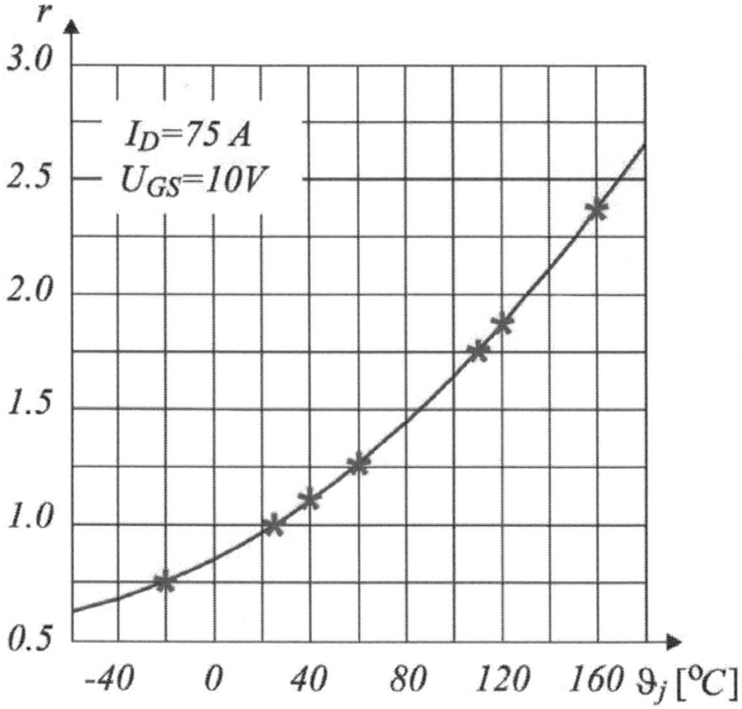

Figure 3. Resistance RDS–ON as a function of junction temperature ϑj.

Table 1. Points extracted from the data-sheet.

ϑ_j [°C]	−20	25	40	60	110	120	160
r	0.750	1.000	1.120	1.250	1.750	1.875	2.375
$R_{DS-ON}(\vartheta_j)[m\Omega]$	2.775	3.700	4.144	4.625	6.675	6.938	8.788

Table 2. Temperature characteristic's coefficients.

Coefficients	K_0	K_1	K_2
	2.61×10^{-5}	5.36×10^{-3}	8.49×10^{-1}

Estimation of Junction Temperature

Unfortunately, the MOS-FET junction temperature is impossible to measure directly and can only be evaluated by using the MOS-FET's thermal model and by the calculated power dissipation. In general, the MOS-FET power dissipation Pdconsists of two main parts as follows:

$$Pd=Pcon+Psw \quad (4)$$

where Pcon represents the conductive and Psw represents the switching losses. MOS-FET power dissipation presents the power sources with an equivalent thermal circuit, which causes ϑj to increase, as well asits influences on RDS−ON.

The nearest point, where temperature measuring can be performed, is the transistor case or heat sink, as indicated in Figure 4a. This can be estimated by using the transistor's simple thermal equivalent circuit, as shown in Figure 4b. Due to the thermal resistances, the junction temperature ϑj is higher than the case temperature ϑC, and ϑC is higher than the heat sink temperature ϑS. The junction and case temperatures can be calculated as follows:

$$\vartheta j=\vartheta C+Pd\cdot R\vartheta j-C \quad (5)$$

and:

$$\vartheta C=\vartheta S+Pd\cdot R\vartheta C-S \quad (6)$$

which leads to:

$$\vartheta j=\vartheta S+Pd(R\vartheta j-C+R\vartheta C-S) \quad (7)$$

where the MOS-FET power dissipation is indicated by Pd, Rϑj−C represents the junction to case thermal resistance for the chosen TO-220 package (Rϑj−c=0.4[oC/W], data-sheet) and RϑC−S is the case to heat sink thermal resistance that must be measured for any particular case, because it depends on different parameters. The case to heat sink resistance (RϑC−S) is usually calculated from MOS-FET and heat sink data-sheets, but accurate data are rarely available, especially when isolation (mica) is placed between the MOS-FET case and heat sink. Isolation usually has a significant impact on RϑC–S resistance. It is difficult to mount the temperature sensor on the MOS-FET case, and the easiest solution is to mount it on the heat sink. These values are never completely accurate because of the lack of a model, namely that it does not describe temperature variations over the range of one switching period. Both Rϑj–C and RϑC–S values cause uncertainty. It is convenient to perform off-line measurement of this parameter. Figure 5 shows the measurements' setup.

The MOS-FET was clamped via a spring in the setup device. The transistor was connected to the power supply by appropriate wires. The transistor current was set at approximately 50 A. The MOS-FET case was painted white in order to improve the conditions for the thermal camera measurement. Painted surfaces had a high radiation coefficient (approximately 0.95). The non-painted MOS-FET case has a smaller radiation coefficient, because of case-surface oxidation, and also, its resistance varies widely, causing large measurement errors.

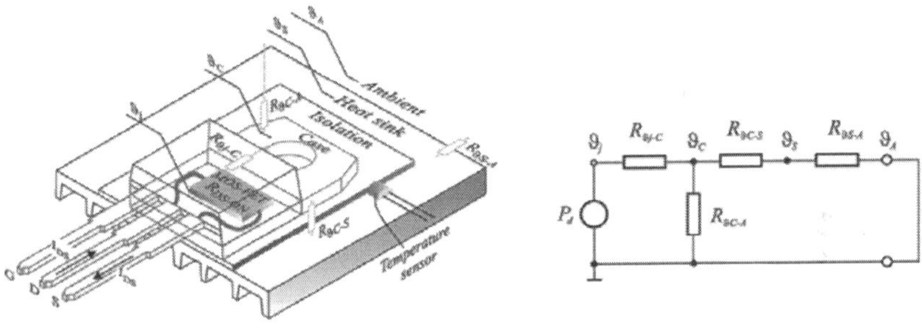

(a) (b)

Figure 4. The MOS-FET TO-220 package; (**a**) thermal components inside the TO-220 package; (**b**) equivalent thermal circuit.

Figure 5. Measurement of Rϑj–S for the MOS-FET TO-220 package.

Using the thermal-camera, the temperatures are measured and marked at three points, as indicated in Figure 5. The warmest are MOS-FET terminals (71.2 ∘C), but the relevant temperatures are the case temperature (53.4 ∘C) and heat sink temperature (35.9 ∘C). The thermal measurement results and corresponding calculated values are indicated in Table 3. Based on the measurements, the thermal resistance R϶j−S is evaluated for any particular case. According to the measured current IDS and voltage UDS (Table 3), the MOS-FET RDS−ON=4.56mΩresistance is calculated, and by rearrangingEquation (3) into:

$$\vartheta_j = \frac{-K_1 + \sqrt{K_1^2 - 4K_0\left(K_2 - r\right)}}{2K_0} \tag{8}$$

it is possible to evaluate the MOS-FET junction temperature, which is also indicated in Table 3. Finally, the measured (evaluated) thermal resistance R϶j−S is indicated in the last column of thesame Table.

Table 3. Temperature characteristics' coefficients.

I (A)	U_{DS} (mV)	ϑ_c (°C)	ϑ_s (°C)	$P = U_{DS} \cdot I$ (W)	$R_{DS-ON} = \frac{U_{DS}}{I}$ (mΩ)	$\vartheta_j = f\left(R_{DS-ON}\right)$ (°C)	$R_{\vartheta j-S}$ (°C/W)
45.2	206	53.4	35.9	9.31	4.56	58.6	2.43

Estimation of MOS-FET Power Dissipation

As can be seen from Equation (4), Pd contains two parts; the conductive Pcon and switching Psw losses. Conductive losses are deterministic for small currents and can be calculated as follows:

$$P_{con} = U_{DS} \cdot I_{DS} \cdot \delta \tag{9}$$

where UDS is transistor voltage, when the transistor is conducting the current IDS, and δ is a duty cycle ratio. At higher currents, the MOS-FET's terminals and bonding wires also heat-up significantly (they are insulated within plastic housing). The interesting fact is also that the current value IDS is unknown during power dissipation calculation, because the value itself is a result of a measurement algorithm. Therefore, in order to evaluate Pcon, it is necessary to use the current measured over previous switching period, which causes the inaccuracy of the measurement procedure. The second part of the losses are the switching losses, which are even harder to evaluate. Different equations exist, but they do not consider all of the effects within the measurement circuit. In a real circuit, the parasitic inductances and capacitances of tracks and elements cause oscillations in the currents and voltages, and this results in some additional switching losses.Figure 6 shows these inductances and capacitances on the boost converter circuit example. As switching losses calculation is always an approximation, they were measured with an oscilloscope. During MOS-FET turn-on and turn-off, the currents and voltages on the terminals were measured simultaneously. Figure 7 shows waveforms during MOS-FET turn-off at the current of 10 A and supply voltage of 10 V. The waveforms were recorded with a Tektronix oscilloscope, that has a bandwidth of 100 MHz, and the current probe was Tektronix TCP305A with amplifier TCPA300. The probe's bandwidth is 50 MHz, and the signal propagation delay is 19 ns. The problem also occurs in the UDS voltage measurement, that is inaccurate because of the parasitic inductances of terminals and bonding wires in the transistor. The switching losses were calculated from the acquired waveforms, by using:

$$P_{sw\ ON/OFF} = \frac{1}{T_s} \underbrace{\sum_{n=1}^{n_{ON/OFF}} u_n \cdot i_n \cdot \Delta t_n}_{W_{ON/OFF}} \tag{10}$$

where un and in are the instantaneously-measured voltage and current values over equidistant time intervals of Δtn. Index ON/OFF suggests that the equation can be used in turn-on or turn-off cases. The data were obtained from the oscilloscope in the form of a CSV (Comma Separated Value) file and transferred to MATLAB for further processing. The calculation was performed under the MATLAB framework, where the current signal delay had also to be accounted for by moving the waveform left, considering the propagation delay of 19 ns. In Figure 7b,d, the instantaneous losses' waveforms' are shown during the switching on and off. The total value of the switching losses was calculated by Equation (10). This procedure was used to measure the turn-on and turn-off losses at three different currents 10,20and30 A, as collected in Table 4. Figure 8 shows the switching losses depending on current IDS (Psw=Psw(IDS)), when the voltage is kept constant at 10 V. It was approximated by a second order polynomial, calculated from the measurement results from Table 4 as follows:

$$P_{sw} = 4.6 \cdot 10^{-4} I_{DS}^2 + 7.2 \cdot 10^{-2} I_{DS} \tag{11}$$

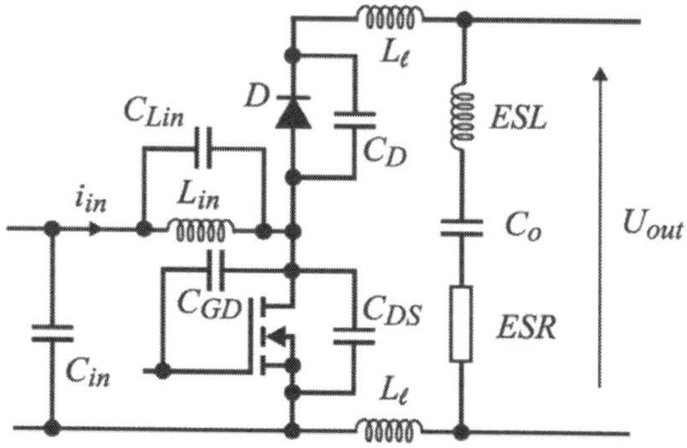

Figure 6. Parasitic elements in the boost converter.

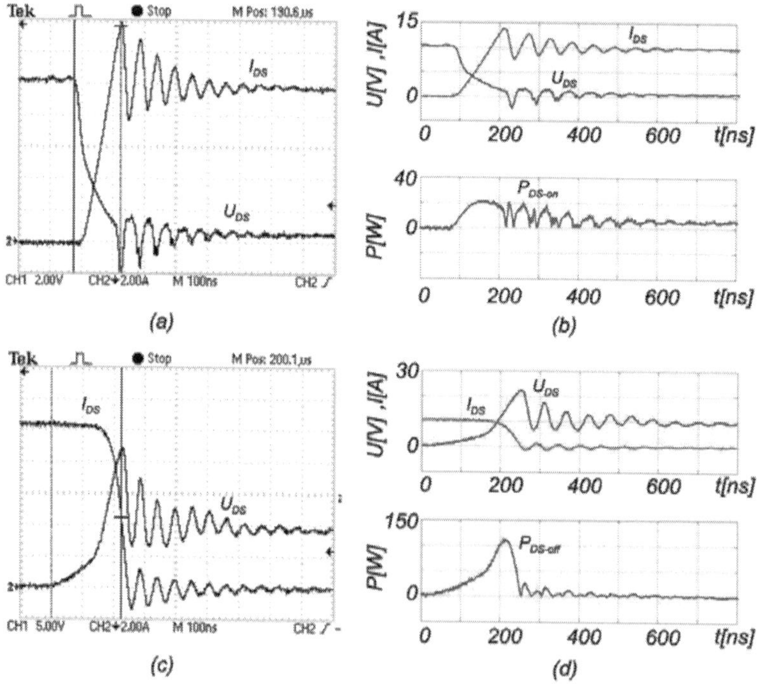

Figure 7. MOS-FET's turn-on/-off waveforms; (**a**) oscillogram of voltage UDS and current ID during the switching-on; (**b**) oscillogram exported to the MATLAB frame for the calculation of switching on dissipation; (**c**) oscillogram of voltage UDS and current IDS during the switching off; (**d**) oscillogram exported to the MATLAB frame for calculation of switching off dissipation.

Table 4. Calculated switching losses by using Equation (10).

I_{DS} (A)	W_{ON} (µJ)	W_{OFF} (µJ)	W_{sw} (µJ)	P_{sw} (W)
10	2.06	9.70	11.80	0.118
20	4.13	28.70	32.80	0.328
30	7.11	55.90	63.00	0.630

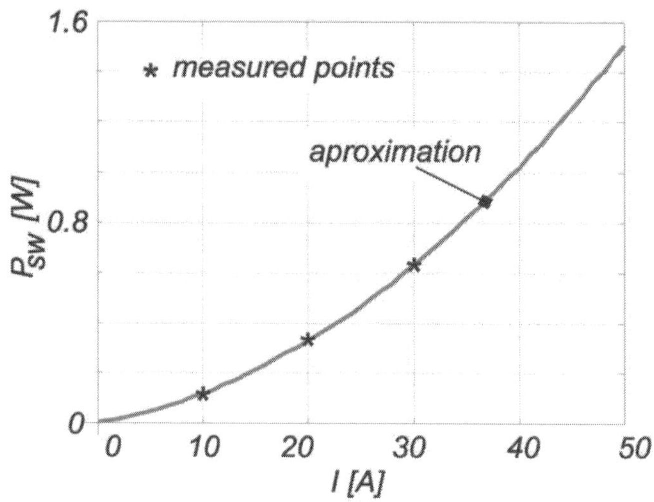

Figure 8. The MOS-FET's switching losses.

The constant supply voltage is assumed for power converters. When powered from batteries, the voltage can vary, but within a narrow range.

Bandwidth Evaluation for the UDS Measurement Chain

The bandwidth of the complete measurement chain (Figure 9) needs to be considered. The targeted bandwidth of the method must be at least equal to the one provided by the use of classical magnetic sensors. Figure 10 shows the block scheme of measurement system, which consists of two frequency-dependent analog circuits. The first one represents differential amplifier with open-loop gain characteristic, as shown in Figure 11 [15] (black line), and the second one represents the switching noise filter (blue line). The frequency characteristics can be calculated as presented in [16]. According to the proposed differential amplifier scheme shown in Equation (9), the equation to be applied is:

$$G_1(j\omega) = \frac{R_6}{R_2} \frac{1}{1 + \left(1 + \frac{R_6}{R_2}\right) \frac{1}{A(j\omega)}} \tag{12}$$

where the open-loop gain A(jω) is defined as:

$$A(j\omega) \cong \frac{A_o}{1 + \frac{j\omega}{\omega_t / A_o}} \tag{13}$$

and Ao is the open-loop gain of amplifier A1 (Figure 9) at the frequency ω=0rad/s and has a very high value (infinity in the ideal case) and ωt represents the cross-over frequency at unity open-loop gain (|A(jω)|=1or0 dB). It must be emphasized that Equation (12) is correct only when R6/R2=R3/R1 (Figure 9). By performing an analysis using Equations (12) and (13), the closed-loop frequency characteristic can be evaluated as:

$$G_1(j\omega) \cong \frac{G_o}{1 + \frac{j\omega}{\omega_b}} \tag{14}$$

where ωb=ωt/(1+R6/R2) and Go=R6/R2 represents the closed-loop gain at the frequency of zero. The quantitative assessments of the bandwidth using Equation (14) for the differential amplifier of different Go (switches S1 and S2 in Figure 9can be in three positions) are indicated in Table 5. The three calculated closed-loop frequency characteristics are featured in the Figure 11 (red lines).

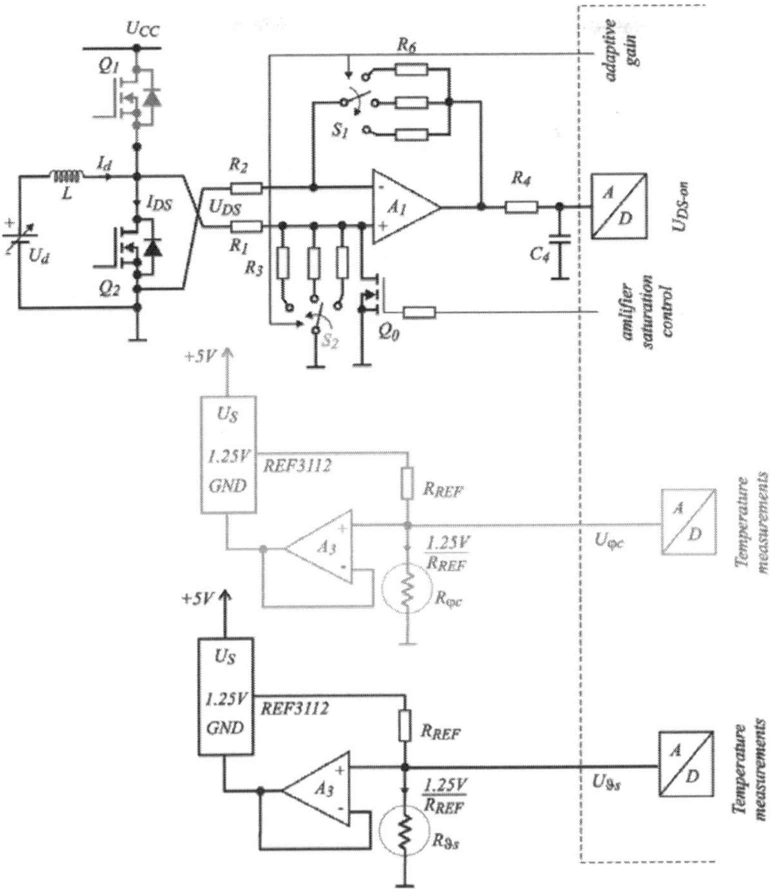

Figure 9. Differential amplifier scheme for UDSmeasurement and the temperature sensors' schemes.

Table 5. Quantitative assessment of closed-loop differential amplifier gain and bandwidth.

Switches S_1 and S_2 Position	R_6 $k\Omega$	R_2 $k\Omega$	$\lvert G_o \rvert$ (dB) $= 20\log(R_6/R_2)$	Bandwidth (MHz) $f_b = \omega_b/(2\pi)$
1	133% ±1%	4.8% ±1%	29	4.0
2	82% ±1%	4.8% ±1%	24	6.4
3	39% ±1%	4.8% ±1%	18	12.4

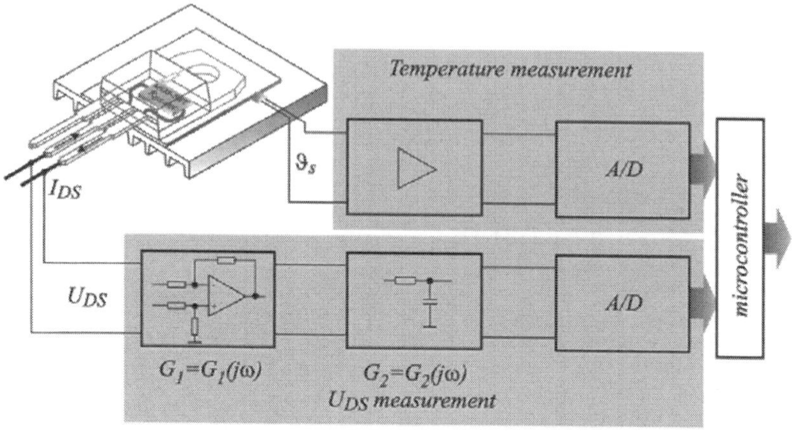

Figure 10. Block scheme of voltage UDS and sink temperature ϑs measurement chains.

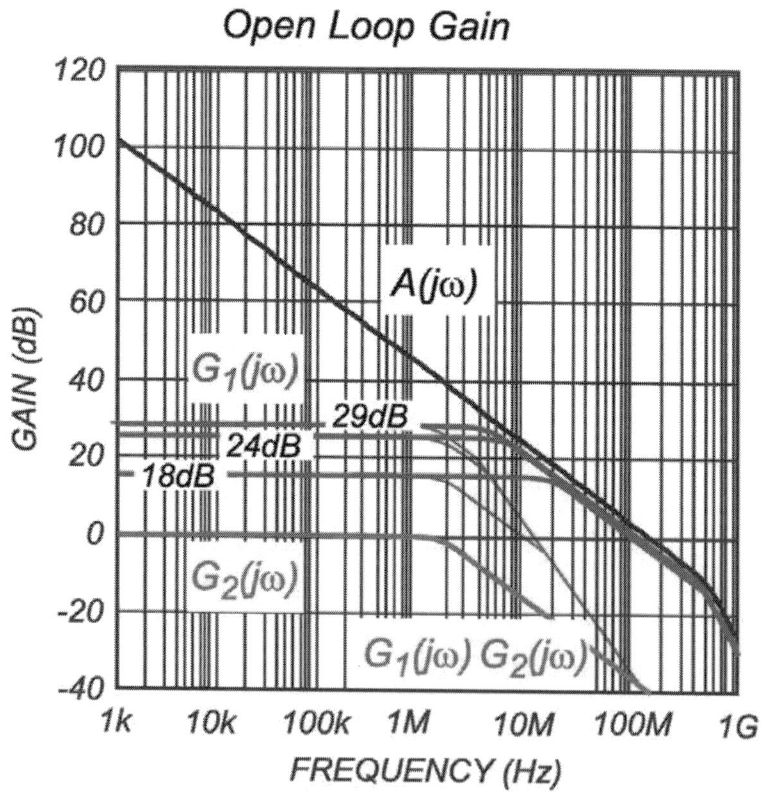

Figure 11. Frequency characteristic of open-loop gain for LMH6611 (black line), closed-loop gains (red lines) and whole measurement chain (tiny green lines).

Further, due to the "hard" switching operation of the observed MOS-FET, the switching noise can disturb the A/D conversion. For this reason, a first order passive filter, which consists of R4 and C4, was introduced. The frequency characteristics of this filter are given as:

$$G_2(j\omega) \cong \frac{1}{1+\frac{j\omega}{\omega_{RC}}} \tag{15}$$

where ωRC=R4C4. According to the chosen values of R4=200Ω and C4=390pF, the bandwidth of this filter was evaluated as ωRC, ($f_{RC}=\omega_{RC}/(2\pi)\cong 2$MHz). The frequency characteristics of the switching noise filter is also featured inFigure 11 (blue line). The complete transfer function, which can be obtained by multiplying G1 and G2, is also presented in the Figure 11 by thin green lines. Therefore, it can be concluded that the bandwidth for such an organized shunt-measurement principle is limited to 2 MHz, which is definitely a higher value than the one that can be reached by using the ordinary magnetic principle.

Due to the component tolerances, the response of the actual differential amplifier, which is the crucial part of the system, will deviate from the ideal one. As a means for predicting the extent of this deviation, the sensitivity analysis was performed. The sensitivities can be quantified using the classical sensitivity function Syx, defined as:

$$S_x^y = \frac{\partial y}{\partial x}\frac{x}{y} \tag{16}$$

where *x* denotes the value of component (according to Equation (14), a gain Go, a frequency ωb), and *y* denotes the circuit parameter of interest (according to Equation (14), the closed-loop gain in absolute form |G1(jω)|). The expression Equation (14) has to be written in its absolute form as follows:

$$G_A = |G_1(j\omega)| = \frac{G_o}{\sqrt{1+(\frac{\omega}{\omega_b})^2}} \tag{17}$$

Thus, the two sensitivities ($S_{G_o}^{G_A}$ and $S_{\omega_b}^{G_A}$) would be evaluated as follows:

$$S_{G_o}^{G_A} = \frac{dG_A}{dG_o}\frac{G_o}{G_A} = 1 \tag{18}$$

$$S_{\omega_b}^{G_A} = \frac{dG_A}{d\omega_b}\frac{\omega_b}{G_A} = \frac{\omega^2}{\omega_b^2 + \omega^2} \tag{19}$$

Low-frequency gain Go depends on resistances R6 and R2; if both have the precision of ±1%, the Go changes for ±2%. Therefore, after using Equation (18), the low-frequency gain changes cause the following:

$$\frac{dG_A}{G_A} = S_{G_o}^{G_A}\frac{dG_o}{G_o} = \pm 2\% \tag{20}$$

A similar analysis can be performed by using Equation (19). This sensitivity is frequency dependent, so two cases will be considered, the first one when the frequency ω=0 and second the one when ω=ωb. Thereby:

$$S_{\omega_b}^{G_A}\bigg|_{\omega=0} = \frac{\omega^2}{\omega_b^2 + \omega^2} = 0 \tag{21}$$

and:

$$S_{\omega_b}^{G_A}\bigg|_{\omega=\omega_b} = \frac{\omega^2}{\omega_b^2 + \omega^2} = \frac{1}{2} \tag{22}$$

Using the sensitivity evaluated above, the variation of the closed-loop gain GAcan be evaluated by:

$$\frac{dG_A}{\omega_b}\bigg|_{\omega=0} = S_{\omega_b}^{G_A}\frac{d\omega_b}{\omega_b} = 0 \tag{23}$$

and

$$\frac{dG_A}{G_A}\bigg|_{\omega=\omega_b} = S_{\omega_b}^{G_A}\frac{d\omega_b}{\omega_b} = \pm 0.95\% \tag{24}$$

Thus, by performing a complete sensitivity analysis, it can be concluded that the parameter (resistances) changes cause the tolerance of the closed-loop gain GA to be in the range of ±2%.

UDS Measurement: Anti-Saturation Circuit

As can be concluded from the previous analysis of the MOS-FET ON resistance that it represents a small quantity due to the fact that the voltage UDS (maximum 300 mV) is also of a small value contaminated by switching noise. Figure 9 shows the scheme of the signal acquisition measuring circuit. The noise represents a significant part of a signal, so the usage of a differential amplifier will improve the signal-to-noise ratio. The operational amplifier is saturated when the on-state time ofQ1 is much longer than the Q2. At the sampling time, when Q2 is ON for a short time, the amplifier output cannot be set to the correct value (due to the operational amplifier bandwidth), and the sensed voltage value contains an error. In order to cope with the problem, a small MOS-FET Q0 was added to the circuit, which is triggered by the same signal as Q1. In order to achieve the highest A/D resolution, adaptive gain was introduced by using switches S1 and S2. The measurements of UDS and temperature ϑs are performed as follows from the measurement scheme presented in Figure 9. In order to test the current measurement chain, the current Id(flowing through inductor *L*) was set to 5 A, and the expected measured voltage can be calculated as:

$$U = I_d \cdot R_{DS-ON} \cdot G = 5\ \mathrm{A} \cdot 3.7\ \mathrm{m\Omega} \cdot 8.15 = 150.8\ \mathrm{mV} \qquad (25)$$

However, during the actual measurement, a voltage of 232 mV was measured, which is different than those calculated and expected (150.8 mV). The low duty cycle on Q2 causes the Q1 connect voltage UCC on the instrumental operational amplifier (A1) input, as indicated in Figure 9, and the amplifier becomes saturated. Due to limited band width, the amplifier output does not reach the steady state in the middle of the Ton interval, because UDS goes from saturation to the right value, as shown in Figure 12a,b. Therefore, when A/D was triggered, the UDS voltage was not the right one. Afterwards, the same test was performed when Q0 was activated, and the results are presented in Figure 12c,d. The performance index is indicated inTable 6. MOS-FET Q0 clearly prevents the amplifier reaching saturation during low duty cycles (when Q1 is switched on for a longer time interval).

Table 6. UDS−ON, instrumental amplifier and anti-saturation effect.

Calculated Voltage (mV)	Measured Voltage (mV) (Q_0 is OFF)	Measured Voltage (mV) (Q_0 is ON)
150.8	232.0	152.0

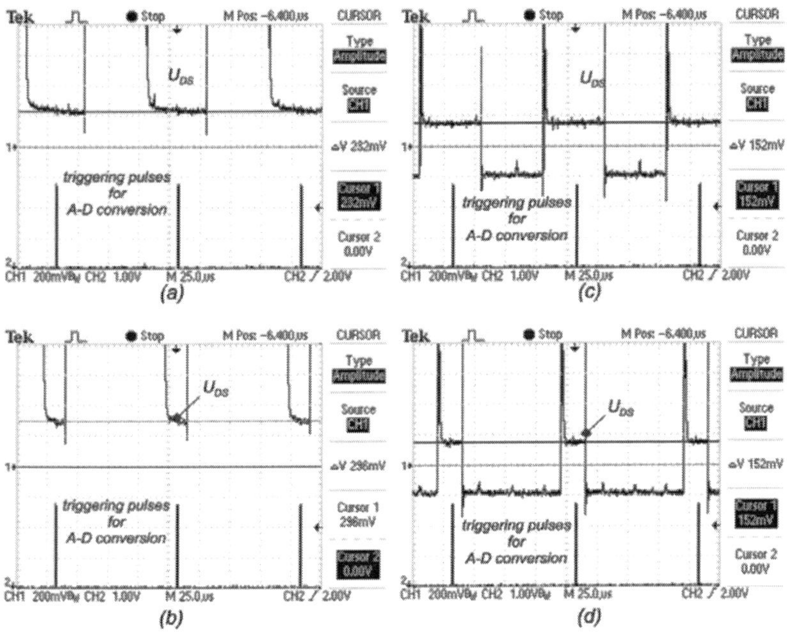

Figure 12. Measurement of UDS.

EXPERIMENTAL SETUP

A test circuit was designed for experimental purposes. It also includes the Hall sensor (LEM-HTB100p) in order to verify the proposed algorithm. A program was written for the ARM microcontroller STM32F407, which controls the MOS-FETs and captures UDS and heat sink temperature ϑS during circuit operation. The MOS-FET dissipation, junction temperature, transistor current IDS and, consequently, the average value of inductor current Id, av are calculated during every switching period. The switching frequency was 10 kHz. Serial communication makes it possible to transfer the calculated data to a PC and to oversee the circuit operation.

PERFORMED MEASUREMENTS, RESULTS AND DISCUSSION

The power source supply unit PS-3010 was used for all tests. It is rated for voltages 0–30 V and a current range of 0–10 A. During verification of the UDS measurement amplifier under a small duty cycle (the δ-duty cycle was less than 20%), the PS-3010 could provide more than its rated current for a short time (up to 40 A). An inductor of 216 μH was used in the converter circuit. The current ripple ΔI at a switching frequency of 10 kHz was less than 1 A. The MOS-FET current measurement was performed by measuring and calculating in the next steps:

1. synchronized measuring of UDS when the transistor is switched on;
2. measuring of heat sink temperature ϑS;
3. using the current from the previous measurement to calculate the power dissipation Pd=Psw+Pcon, where the switching losses are evaluated using formula Psw=4.6·10−4I2d,av+7.2·10−2Id,av and conductive losses are evaluated as Pcon=U2DSRDS−ON;
4. junction temperature is evaluated using ϑj=ϑS+Pd(Rϑj−C+RϑC−S); for initial measurement, there is no previous current IDS data, so ϑj is considered equal to ϑS.
5. based on estimated ϑj, the R∗DS−ON is recalculated;
6. the real current I∗d,av is calculated using measured UDS and evaluated R∗DS−ON.

In order to verify the obtained results and for the comparison between the reference measurement system based on the magnetic Hall sensor, the proposed measurement procedure was performed based on the UDS measurement. The duty cycle (δ=Ton/T) was modulated from 0.05–0.3. Both measurements were performed; Id,av was measured using the proposed algorithm (UDS and ϑS measurement) and Iref using the Hall sensor. the obtained results are shown in Table 7. Figure 13 shows the static characteristic of reference and real current measurements. Significant relative error ε appears when applying a relatively small duty cycle (0.05<δ<0.1). This is expected, as, due to the relatively low-pass bandwidth of the operational amplifier, the anti-saturation circuit is incapable of setting the output of the instrumental amplifier at the correct value in a very short time [17].

Table 7. Verifications of the proposed measurement algorithm.

Coefficients	Error without Compensation	Error with Compensation
sine wave	10.2%	< 2%
square wave	13.2%	< 2%

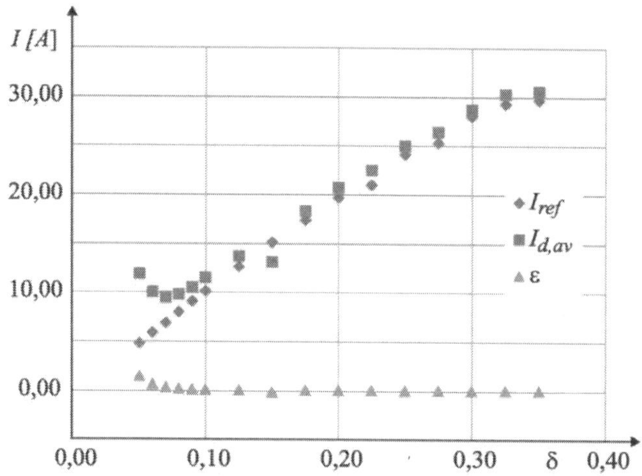

Figure 13. The static characteristics of reference measurement system and the proposed measurement algorithm.

Furthermore, the proposed algorithm was also tested for the sine wave and square modulation (0.1<δ<0.3). The raw measurement results were acquired in real time by a microcomputer and then processed in MATLAB. The maximal errors (Id, av compared with Iref) were between 10.2% and 13.2% and are extracted, respectively, from Figure 14a,b, where the obtained results are presented. The voltage UDS and heat sink temperature are also acquired and displayed in Figure 14a,b. As was expected from previous verification of the proposed algorithm, the error appeared because of the poor performance of the sensor at low duty cycles. The acquisition circuit is not fast enough to track UDS during the short conduction time of Q2. Calibration based on the reference current measurement by the Hall sensor was performed in order to compensate for this error. It was measured for some small values of the duty cycle. The static characteristic ε=ε (Id, av, Iref) was approximated by the function:

$$\varepsilon(\delta)=\text{Id,av}-\text{IrefIref}=a(\delta-b)2+c \tag{26}$$

where coefficients a, b and c are indicated in Table 8. The order of the function and coefficients was determined experimentally. By the help of Equation (26), the measured results were compensated. The performed algorithm is calibrated with the reference current measurement. The obtained results are shown in Figure 14c, d. The performance index without and with compensating approaches is indicated in Table 9. The voltage UDS and heat sink temperature are also acquired and displayed inFigure 14c, d.

Table 8. Error characteristic's coefficients.

Coefficients	a	b	c
–	580	30	0.02

Table 9. Error characteristics' coefficients.

Coefficients	Error without Compensation	Error with Compensation
sine wave	10.2%	< 2%
square wave	13.2%	< 2%

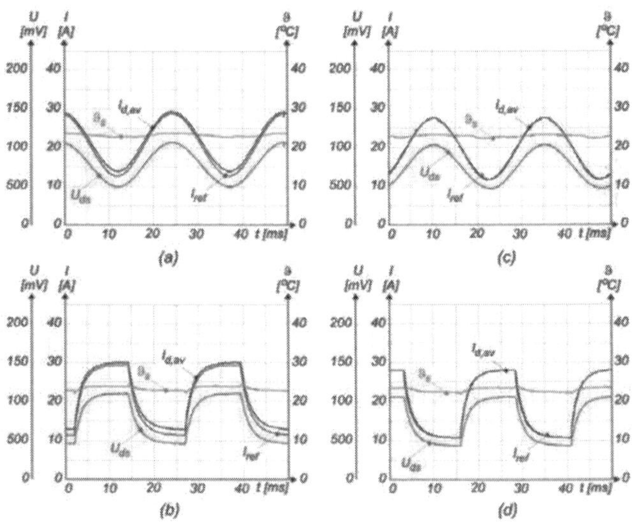

Figure 14. Sine wave and square modulation when δ∈ (0.1–0.3); (a) measured UDS, ϑs, Iref and calculated Id,av without compensation (sinus modulation); (b) measured UDS, ϑs, Iref and calculated Id,av without compensation (square modulation); (c) measured UDS, ϑs, Iref and calculated Id,av with compensation (sinus modulation); (d) measured UDS, ϑs, Iref and calculated Id,av with compensation (square modulation).

CONCLUSIONS

Based on the obtained results, it is possible to claim that MOS-FETs can be used as sensors in power electronics' converters. Clean signals can be achieved, with careful measurement circuit design, which are comparable to other sensor types. This sensor has many advantages; the measurement principle does not add many additional losses to the converter circuit, thus eliminating the offset problem of the Hall sensors. The main obstacle for this measurement principle is the changes of RDS−ON during the operation at small duty cycles and due to the junction temperature. The temperature dependence of RDS−ON must be considered. Approximation with potential function was proposed for limited accuracy at small duty cycles. As a testing platform, the step-up converter was chosen, and the obtained results are promising.

AUTHOR CONTRIBUTIONS

The work presented in the paper was initiated by B.P. from the company Piktronik. R.P. designed all the presented electronics circuits and wrote the software under the guidance of M.M., B.P. and M.R. The theoretical analysis was done by M.M. and M.R.

REFERENCES

1. Wrzecionko, B.; Steinmann, L.; Kolar, J.W. High-bandwidth high-temperature (250 °C/500 °F) isolated DC and AC current measurement: Bidirectionally saturated current transformer. *IEEE Trans. Power Electron.***2013**, *28*, 5404–5412.
2. Lee, W.-C.; Hyun, D.-S.; Lee, T.-K. Novel control method for threephase PWM rectifiers using a single current sensor. *IEEE Trans. Power Electron.***2000**, *15*, 861–870.
3. Zidat, F.; Lecointe, J.; Morganti, F.; Brudny, J.; Jacq, T.; Streiff, F. Non invasive sensors for monitoring the efficiency of AC electrical rotating machines. *Sensors* **2010**, *10*, 7874–7895.
4. Reyes Barranca, M.; Mendoza-Acevedo, S.; Flores-Nava, L.; Avila-Garcia, A.; Vazquez-Acosta, E.; Moreno-Cadenas, J.; Casados-Cruz, G. Using a floating-gate MOS transistor as a transducer in a MEMS gas sensing system.*Sensors* **2010**, *10*, 10413–10434.
5. Dai, C.; Tai, Y.; Kao, P. Modeling and fabrication of micro FET pressure sensor with circuits. *Sensors* **2007**, *7*, 3386–3398.

6. Dai, C.; Lu, P.; Wu, C.; Chang, C. Fabrication of wireless micro pressure sensor using the CMOS process. *Sensors* **2009**, *9*, 8748–8760.
7. Kim, H.R.; Jahns, T.M. Phase current reconstruction for ac motor drives using a dc-link single current sensor and measurement voltage vectors. *IEEE Trans. Power Electron.* **2006**, *21*, 1413–1419.
8. Cusido, J.; Romeral, L.; Ortega, J.; Garcia, A.; Riba, J. Signal injection as a fault detection technique. *Sensors* **2011**, *11*, 3356–3380.
9. Aiello, O.; Fiori, F. A new mirroring circuit for power MOS current sensing highly immune to EMI. *Sensors* **2013**, *13*, 1856–1871.
10. Lavrič, H.; Fišer, R. Lossless current sensing technique on MOSFET RDS (on) with improved accuracy. *Electronics Letters* **2010**, *46*, 370–371.
11. Chen, H.; Pickert, V.; Atkinson, D.J.; Pritchard, L.S. On-line monitoring of the MOS-FET device junction temperature by computation of the threshold voltage. In Proceedings of the 3rd IET International Conference Power Electronics, Machines and Drives (PEMD 2006), Dublin, Ireland, 4–6 April 2006; pp. 440–444.
12. Musallam, M.; Acarnley, P.P.; Johnson, M.; Pritchard, L. Real-time power electronic device junction temperature estimation. In Proceedings of the Second International Conference on Power Electronics, Machines and Drives (PEMD 2004), Edinburgh, UK, 31 March–2 April 2004; pp. 231–235.
13. International Rectifier, IRFB4110GPbF: HEXFET Power MOS-FET. International Rectifier, 2009. Available online: http://www.irf.com/product-info/datasheets/data/irfb4110pbf.pdf (accessed on 17 July 2015).
14. Buttay, C.; Bergogne, D.; Herve, M.; Bruno, A.; Rene, E.; Pascal, B. Towards a Sensorless Current and Temperature Monitoring in MOS-FET-Based H-Bridge. In Proceedings of the 2003 IEEE 34th Annual Power Electronics Specialist Conference, Acapulco, Mexico, 15–19 June 2003; pp. 901–906.
15. Texas Instruments, LMH6611: Single Supply 345 MHz Rail-to-Rail Output Amplifiers. Texas Instruments, 2013. Available online: http://www.ti.com/lit/ds/symlink/lmh6611.pdf (accessed on 17 July 2015).
16. Sedra, A.S.; Smith, K.C. *Microelectronic Circuits*; Oxford Series in Electrical & Computer Engineering; Oxford University Press: Oxford, UK, 2009.
17. Pajer, R. Current Sensing on Power MOS-FET in Electrical Motor Drives. Masters Thesis, University of Maribor, Maribor, Slovenia, 2014.

CITATION

Rok Pajer , Miro Milanovič , Branko Premzel and Miran Rodič, MOS-FET as a Current Sensor in Power Electronics Converters, doi:10.3390/s150818061http://www.mdpi.com/1424-8220/15/8/18061/htm

CHAPTER 6

An Image Reconstruction Algorithm for Electrical Capacitance Tomography Based on Simulated Annealing Particle Swarm Optimization

P. Wang[1], *J.S. Lin*[2], *M. Wang*[1]

[1] School of Electric and Control of Xi'an University of Science and Technology, Xi'an, China
[2] Department of Computer Science and Information Engineering at National Chin-Yi University of Technology, Taichung, Taiwan

ABSTRACT

In this paper, we introduce a novel image reconstruction algorithm with Least Squares Support Vector Machines (LS-SVM) and Simulated Annealing Particle Swarm Optimization (APSO), named SAP. This algorithm introduces simulated annealing ideas into Particle Swarm Optimization (PSO), which adopts cooling process functions to replace the inertia weight function and constructs the time variant inertia weight function featured in annealing mechanism. Meanwhile, it employs the APSO procedure to search for the optimized resolution of Electrical Capacitance Tomography (ECT) for image reconstruction. In order to overcome the soft field characteristics of ECT sensitivity field, some image samples with typical flow patterns are chosen for training with LS-SVM. Under the training procedure, the capacitance error caused by the soft field characteristics is predicted, and then is used to construct the fitness function of the particle swarm optimization on basis of the capacitance error. Experimental results demonstrated that the proposed SAP algorithm has a quick convergence rate. Moreover, the proposed SAP outperforms the classic Landweber algorithm and Newton-Raphson algorithm on image reconstruction.

INTRODUCCION

As one of the electrical process tomography imaging technologies, Electrical Capacitance Tomography (ECT) is featured in lower costs, no-irradiative and non-invasive methods, etc., and applicable to the visible measurement of two-phase and multiple-phase flows (York, 2001 and Griffiths, 1988). The principle of ECT can be described as: different objects have different permittivities. If the concentration and the composition of the component phase are changed, the permittivity will change to fit the mixture. Variation in permittivity will cause a change of the capacitance measurements and the capacitance measurements reflect the size and distribution of the medium phase concentration of the mixture. On this basis, using a corresponding image reconstruction algorithm can reconstruct the distribution of the test area of the pipeline. Because ECT is non-linearity and the number of capacitances independently measured are much less than the number of pixels for image reconstruction, there is no resolution for the reverse problem. Furthermore, the sensitivity field of ECT is featured in "soft field", i.e. sensitivity is not evenly distributed, the reverse problem equation is in a seriously abnormal state (Yang, 1997). Therefore, image reconstruction algorithm has been the bottleneck for the further development of ECT, and a high precise image reconstruction algorithm is required.

The existing ECT image reconstruction algorithms can be divided into two mainly types: non-iterative algorithm and iterative algorithm. As one of the typical non-iterative algorithms, Linear Back Projection (LBP) is simple and quick, but unsatisfying in imaging precision. So LBP is only used as a qualification method (Peng, Lu, & Yang, 2004). Iterative methods include: Tikhonov regularization method (Peng, Jiang, Lu, & Xiao, 2007), Landweber algorithm (Yang, Spink, York, & McCann, 1999), Newton-Raphson algorithm (Yang & Peng, 2003) and Conjugate Gradient method, etc. (Wang, Zhu, & Zhang, 2005). Tikhonov method may cause detailed distortion of the reconstructed images due to over-smoothness of regularization functions. As a widely used method in recent years, Landweber returns satisfying results only with large number of iterations as to complex flow patterns. Newton-Raphson algorithm is featured in local convergence, but the iterative convergence can't be guaranteed if the initial value is not selected appropriately. Conjugate gradient method is applicable to positive definite matrix and thus it can't obtain better effects when it is applied to complex flow patterns.

In this paper, we introduce an image reconstruction algorithm with LS-SVM and APSO, which is named as SAP. The proposed SAP is described as follows: firstly, we construct LS-SVM and excise the error between the capacitances arising from sensitivity matrix and the actual capacitance measurements; then based on the error, we constructed the fitness function and simulated annealing mechanism for particle swarm optimization; finally, we search for the optimum solution for image reconstruction with APSO.

ECT SYSTEM

As shown in Figure 1, ECT System is mainly consisted with three units: a capacitance sensor unit, a measurement and data collection unit, and an image reconstruction unit. By utilizing capacitive fringe effect, the sensor can produce a corresponding capacitance for a medium with certain permittivity. The combination of all sensing electrodes may provide multiple capacitance measurements, which can be taken as the projection data for image reconstruction. The capacitance measurement and data collection unit primarily functions as rapidly, stably and accurately measuring minor capacitance. It changes in various arrays of electrode couples, and transmits the acquired data to a computer. This unit is mainly comprised of three modules: a capacitance measurement module, a data collection control module, and a communication module. The capacitance measurement module is used to realize switching of capacitance to voltage (CV), to measure minor capacitance and effectively inhibit stray capacitance. Currently, two of the most mature methods to measure capacitance are as follows: capacitance charge-discharge method and AC-based CV switching circuit (Yang, 1996 and Yang, 2001) The data collection control module generally takes DSP as the control core and takes ADC for data acquisition. Data communication adopts USB2.0 Technology (Yang et al., 2010). As ECT System has more measurement channels, it is difficult for a single DSP to meet real-time requirements. Therefore, CPLD or FPGA is generally adopted to conduct auxiliary control of DSP (Ma, Wang, & Tian, 2006). ECT image reconstruction unit is composed of two parts: hardware and software. Hardware indicates a general-purpose computer, and software indicates image reconstruction algorithm.

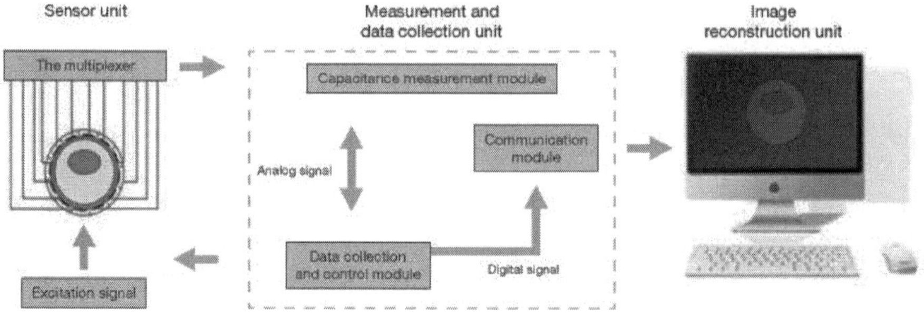

Figure. 1. Constitutes of electrical capacitance tomography system.

ECT IMAGE RECONSTRUCTION

ECT image reconstruction process includes forward and reverse questions to be resolved. As the forward question, capacitance values of all electrode pairs on basis of the permittivity distribution and excitation voltages of the known sensitivity field. The mathematic model of forward question of ECT is expressed as follows (Yang & Peng, 2003):

$$C_{i,j} = \iint_{\Gamma} \varepsilon(x,y) \cdot S_{i,j}(x,y)\,dx\,dy \tag{1}$$

where *Ci,j* is the capacitance between the electrode pair of *i-j*, ′*(x,y)* is the permittivity distribution on cross-section of pipes, *Si,j(x,y)*is the sensitivity functions when the capacitance between electrode pair of *i-j* is distributed on the cross-section of pipe, and G is the electrode surface. It can be seen that the sensitivity of the electrode in a point is related to its position, namely the sensitivity is not evenly distributed within the sensitivity field, which is the so-called effects of "soft field".

The capacitance sensor comprising of n electrodes can provide $M = n(n - 1)/2$independent capacitances. With M equations similar with equation (1), such equations shall be linearized and discredited to get:

$$C = S \cdot G \tag{2}$$

where C is a normalization capacitance vector of M dimension, G is N dimension normalized permittivity distribution vector, i.e. the grey level of pixels for visualization, and S is $M \times N$ factor matrix, reflects influence of medium distribution variation on capacitance C, and is called as sensitivity matrix (Tikhonov & Arsenin, 1977).

$$S_{i,j}(\ell)=\frac{C_{i,j}(\ell)-C^{l}_{i,j}}{C^{h}_{i,j}-C^{l}_{i,j}} \quad (3)$$

Where $Si,j(\ell)$ is the ℓth unit of electrod pair to i-j sensitivity matrix, $C_{i,j}(\ell)$ is the measurement capacitance, $C_{i,j}(1)$ is the capacitances when the sensor is full of lower permittivity mediums, and $Ci,j(h)$ is the capacitances when the sensor is full of higher permittivity mediums.

The reverse question is to reconstruct permittivity distribution diagram in the sensitivity field with capacitance measurements and sensitivity matrix S representing grey levels of pixels. Currently, most of the image reconstruction algorithms are achieved with the basis of:

$$G = S^{-1}C \quad (4)$$

There is no annalitic resolution for equation (4). Firstly, permittivity distribution is not linear with the capacitance measurements at boundaries in sensitivity field. Secondly, the data M is independently measurements and far less than the number of pixels N, therefore the solution of equation (4) is not unique. Furthermore, the equation (4) is ill posed, and the resolution is not stable. Minor error of C may have significant influences on G (Yeung & Ibrahim, 2003.). In addition, the matrix S is not truly constant, but varies with the actual permittivity distribution. Therefore, we try to resolve reverse question of ECT with a heuristic algorithm to achieve high-precise imaging in this paper.

IMAGE RECONSTRUCTION ALGORITHM ON BASIS OF SIMULATED ANNEALING PARTICLE SWARM OPTIMIZATION

Conventional Particle Swarm Optimization Algorithm

Particle swarm optimization (PSO) is a well known heuristic algorithm, which is firstly proposed by Kennedy and Eberhart (1995) and is sourced from studies on food-catching of birds. Many kinds of PSO algorithms have been widely studied and have made certain achievements in the last twenty years (Rezazadeh et al., 2009, Arce et al., 2012 and Gerardo et al., 2009). In PSO system, each alternative resolution is called as a "particle". Particles are co-existing and shall be optimized. That is because each particle should "fly" towards to a better position in the question space according its own experiences to explore the best resolution. The mathematic expression of PSO is shown as follows (Eberhart & Shi, 2000).

In this paper, we presume the space is *D*-dimension and the scale of particles are *m*. The position of the *ith* particle is $Xi = (xi_1, xi_2, \ldots, xi_D)$. The best position of the *ith* particle in the "flying" history is $Pi = (pi_1, pi_2, \ldots, pi_D)$, and we presume the best value of $Pi(i = 1,2,\ldots,m)$ is located at Pg; the velocity of the *ith* particle is the vector of $Vi = (vi_1, vi_2, \ldots, vi_D)$; the position of *ith* particle will change according to the following equations:

$$v_{id}(t+1) = \omega \cdot v_{id}(t) + c_1 \cdot r_1 \cdot [P_{id}(t) - x_{id}(t)] + c_2 \cdot r_2 \cdot [P_{gd}(t) - x_{id}(t)] \tag{5}$$

$$x_{id}(t+1) = x_{id}(t) + v_{id}(t+1) \tag{6}$$

Where c_1 and c_2 are positive constants and called as speedup factors; r_1 and r_2 are two random numbers between [0, 1]; w is called as the inertia factor; i is the *ith* particle $1 \leq i \leq m$, d is the *dth* dimension of each particle $1 \leq d \leq D$. The initial position and speed of particle swarm is generated randomly, and then iterated according to equations (5) and (6). The position variance range is $[xd_{,min}, xd_{,max}]$ and the speed variance range is $[vd_{,min}, vd_{,max}]$. The boundary value shall be taken if the dimension of *xid* or *vid* exceeds the boundary.

A Simulated Annealing Particle Swarm Optimization Algorithm

The inertia factor v in equation (5) keeps particles with the movement inertia of the last generation, and thus particles intend to extend the search range. When v is larger, the last speed has significant influence and thus has better overall search capabilities; otherwise, the last speed has less influence and thus the local search capability is strong. Local optimization can be skipped by dynamically adjusting v. In their studies, Shi and Ebethartln pointed out that the inertia factors reduce gradually with numbers of iteration and thus questions caused by equivalent factor are improved (Eberhart & Shi, 1998). In other words, if the resolution after movement of particles are inferior to the solution before movement of particles, the movement shall be accepted at a certain possibility, which shall reduce gradually when time passes. However, the study result of Shi and Ebethartln did not definitely give a mathematics definition; they only give a qualitative description with their experiences.

Simulated annealing algorithm is another widely used iterative heuristic algorithm. The powerful feature of its intrinsic hill climbing capability (Kirkpatrick, 1983 and Sait et al., 2013). On basis of the ideas of

Simulated annealing, we introduced a cooling process function (Lin, 2001) to give a quantitative description on the process that the possibility reduces with the reduction of numbers of iteration.

$$T(t)=\frac{1}{\tau+1}\left[\tau+\tanh(\Omega)^{t}\right]T(t-1) \tag{7}$$

$$\begin{aligned} v_{id}(t+1) &= T(t)v_{id}(t)+c_1 r_1\left[P_{id}(t)-x_{id}(t)\right] \\ &+c_2 r_2\left[P_{gd}(t)-x_{id}(t)\right] \end{aligned} \tag{8}$$

Selection of Fitness Functions

For each iteration of PSO, the fitness function is used to determine whether the position of particle is satisfied or dissatisfied. When PSO is applied to the image reconstruction of ECT, the fitness function is generally taken as follows:

$$F=\min\,(C-S\cdot G) \tag{9}$$

Where C is the normalization measurement capacitance vector, S is a sensitivity matrix and G is position vector of particle (i.e. the required permittivity distribution vector).

Due to the matrix S is not truly constant, but varies with the actual permittivity distribution. PSO taking takes equation (9) as the fitness function, which can achieve an overall optimization of convergence and there may be larger error between the optimization resolution and actual distribution of permittivity. The fitness function in this paper is given as the following:

$$F=\min(\|C-S\cdot G\|-\|\Delta C\|) \tag{10}$$

Where $\|\Delta C\|$ is the output when LS-SVM takes C as input. The fitness function takes use of the results predicted by LS-SVM so as to eliminate errors arising from that different flow patterns under the same sensitivity matrix S.

LEAST SQUARES SUPPORT VECTOR MACHINE AND ITS APPLICATIONS IN IMAGE RECONSTRUCTION

Least Squares Support Vector Machine

LS-SVM is a kernel function study machine complying with Structural Risk Minimization (SRM) algorithm of the least square algorithm and principle of SRM (Gu, Zhao, & Wu, 2010). The concept is described as follows: specify sample collection $\{xi,yi\}^{k}{}_{i=1}$ map *n*-dimension input vector *Xi* and output vector *Yi* from the original space *Rn* to the high dimension special space *Rn+* through non-linear transformation $\Phi(x)$. In this space the optimization linear decision function is given as the following:

$$f(\mathbf{x}) = \mathbf{w} \cdot \Phi(\mathbf{x}) + b \tag{11}$$

Where, *W* is a hyper plane weight vector, and *b* is a polarization item. For LS-SVM, the question to be optimized is as follows:

$$\min \Phi(\mathbf{w}, e) = \frac{1}{2}\mathbf{w}^T\mathbf{w} + \frac{1}{2}\gamma\sum_{i=1}^{K} e_i^2 \tag{12}$$

Where, *ei* is an error variant; $\Upsilon > 0$ is a punitive parameter controlling the punishment for samples exceeding errors. LS-SVM converts the inequality constraints into equality constraints (Ott, Grebogi, & Yoke, 1990). For LS-SVM for regression estimate, the constraints are as follows:

$$y_i = \mathbf{w}^T\varphi(\mathbf{x}_i) + b + e_i, \qquad i = 1,2,\ldots,K \tag{13}$$

According to KKT conditions, the Lagrange factor $\alpha_i \in R$ (i.e. support vector factor) is introduced to construct Lagrange function.

$$L(\mathbf{w}, b, \mathbf{e}, \alpha) = \frac{1}{2}\mathbf{w}^T\mathbf{w} + \frac{1}{2}\gamma\sum_{i=1}^{K} e_i^2 - \sum_{i=1}^{K}\alpha_i\left(\mathbf{w}^T\varphi(x_i) + b + e_i - y_i\right) \tag{14}$$

According to the optimization conditions, we make the partial derivative of *L* against *w, b, e* and a as zero, and eliminate variant *w* and *e* and thus get the following linear equation:

$$\begin{bmatrix} 0 & \Pi^T \\ \Pi & \Lambda+\frac{1}{\gamma}I \end{bmatrix}\begin{bmatrix} b \\ \alpha \end{bmatrix}=\begin{bmatrix} 0 \\ y \end{bmatrix} \tag{15}$$

Where, y = [y_1, y_2,…, yk], a = [a_1, a_2,…, ak] and P is column vectors with k dimension as 1 and $\Lambda ij=\varphi(xi)T\varphi(xj)$ According to the principle of Mercer, the mapping function *Φ*() and kernel function *Φ*() existing and thus:

$$K(\mathbf{x}_i,\mathbf{x}_j)=\varphi(\mathbf{x}_i)^T\varphi(\mathbf{x}_j) \qquad i,j=1,2,\ldots,\kappa \tag{16}$$

The function estimates expression of the least square algorithm is as follows:

$$f(\mathbf{x})=\sum_{i=1}^{\kappa}\alpha_i K(\mathbf{x},\mathbf{x}_i)+b \tag{17}$$

Where, ai(i = 1,2,…,k) and b are obtained from equation (15), the specific forms of f(x) is subject to types of function $K(x,xi)$. There are many kinds of kernel functions. In this paper, we take the kernel function of radial basis (i.e. Gaussian) with higher regression capabilities which is defined as follows:

$$K(\mathbf{x},\mathbf{x}_i)=\exp\left(-\frac{\|\mathbf{x}-\mathbf{x}_i\|^2}{\sigma^2}\right) \tag{18}$$

Where, s is Gaussian kernel parameter.

Functions of LS-SVM in ECT Image Reconstruction

As described above, due to the soft field characteristics of ECT, the capacitance calculated through equation (2) and the actual measurement of capacitance have the following e-rrors:

$$\Delta C=C-S\cdot\bar{G} \tag{19}$$

where, $\bar{G}$ is actual permittivity distribution vector. In order to get the capacitance error *ΔC*under any permittivity distribution, we take use of LS-SVM to exercise certain of quantities of samples with LS-SVM. Taking samples of vectors into equation (19) and get:

$$y=\|\Delta C\|=\|C-S\cdot\bar{G}\|=f(x):R^n\rightarrow R^1, \tag{20}$$

where the input vector x is normalization capacitance vector, and the output y is the norm of samples of capacitance error. When carrying out exercises with LS-SVM, the input sample collection is actual capacitance vector under all kinds of flow patterns, and output sample collection is the norm of samples of capacitance error under all kinds of flow patterns.

SIMULATION RESULTS AND ANALYSIS

Algorithm Flow Process

The SAP algorithm comprises of LS-SVM exercise forecast stage and APSO search stage. The algorithm flow process is shown as Figure 2, in which t is current iteration times of APSO.

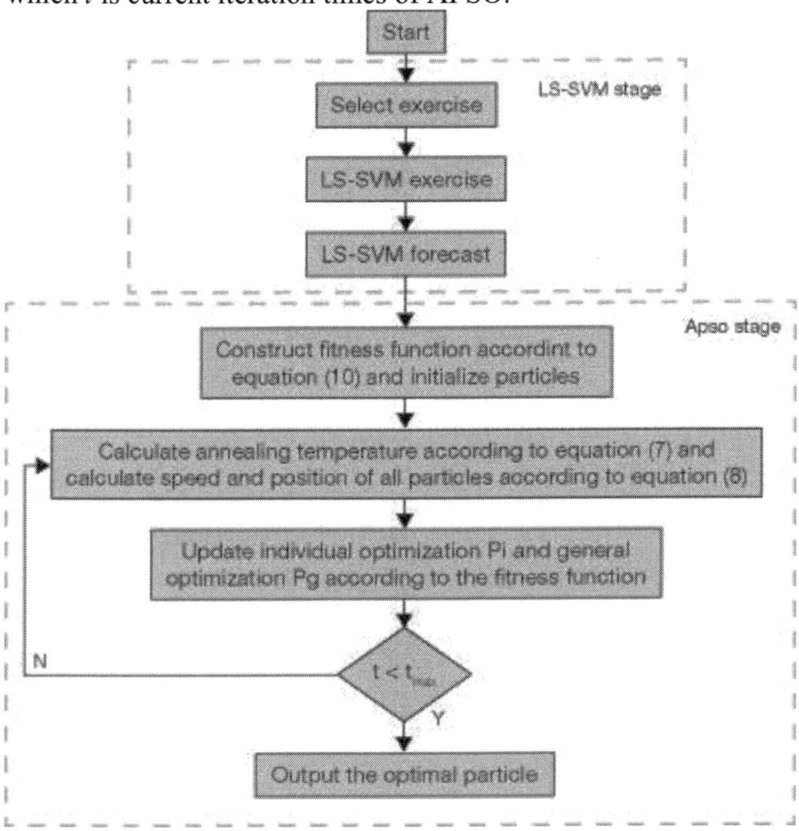

Figure. 2. SAP algorithm process.

In LS-SVM stage, 40 groups of samples including 4 kinds of flow patterns are used as exercise samples, the punitive parameter g and the kernel parameter s shall be selected through experiences. Based on the exercise results of LS-SVM, forecast of the test samples and capacitance error $\|\Delta C\|$ are achieved. In this paper, we select 8 electrode capacitance sensors to get 28 separated capacitance measurements and thus the input sample data of LS-SVM *xi* is *28*-dimension (i = 1,2,...,40). Capacitance measurements can be obtained with finite element methods. In finite subdivision, we take triangle unit to subdivide the imaging area into *800* units, and we take finite subdivision unit as the pixel unit of images and the permittivity distribution $\bar{G}$ under all kinds of flow patterns of sample is an *800*-dimension vector.

In APSO stage, we firstly construct fitness function according to equation (10), then initialize particle swarm. The swarm scale is set as m=40, the number of dimensions is same as the number of pixels in image domain, i.e. D = 800. The maximum iteration number is t_{max}. We set up limit position of particles and variance range between limit speed and particle speed. Finally, we search for the optimum solution for image reconstruction with APSO.

Experimental Results and Evaluation on Algorithm

In order to validate effectiveness of the algorithm, we take SAP algorithm to make image reconstruction for typical flow patterns (i.e. core flow, bubble flow, laminar flow, and circular flow), and then compared them with the imaging results of Newton-Raphson algorithm and Landweber algorithm. The simulation and calculation is carried out with MATIAB 7.10 on Intel Pentium 820 MHz CPU (RAM 512 MB). The experimental results are shown in Table 1. In imaging area, the dark area is even medium of permittivity 40 and the other areas is air (i.e. permittivity 1.0).

Table 1: Imaging result.

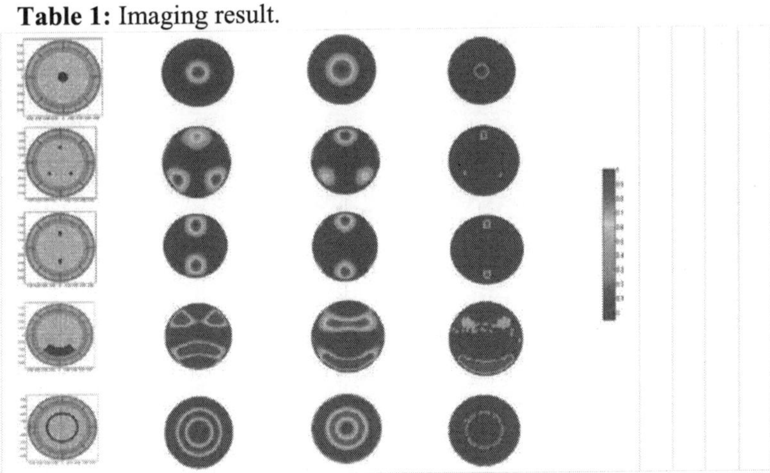

As shown in Table 1, we can see that imaging results with Landweber algorithm and Newton-Raphson algorithm are near to the original; however, there are too many false images. Obviously, the quality of images obtained with SAP is much better, for which the resolution of images is much higher and there is nearly no false image.

When the quality of image is analyzed, the relative image error shall be used as evaluation index of image quality, which is defined as follows:

$$\varepsilon_{image} = \frac{\|\hat{g} - g\|}{\|g\|} \tag{21}$$

where, $\hat{g}$ is permittivity distribution vector obtained with reconstruction algorithm, and g is permittivity distribution vector in the original. i·i is a vector sample norm, which here is taken as 2.

The experimental results of the relative image error are shown in Table 2. From Table 2we can see that the quality of reconstruction image with SAP for all above flow types are significantly improved by comparing with Newton-Raphson and Landweber algorithms.

Table 2. Relative image error.

Original	Newton-Raphson	Landweber	SAP
1	33%	41.80%	16.00%
2	37.60%	40.20%	19.50%
3	42%	40%	21%
4	71%	65%	41%
5	32.80%	40.10%	24%

The elapsed time required for reconstruction of the three different algorithms are shown in Table 3. Obviously, the number of iterations for landweber and Newton-Raphson algorithms are greater than that for SAP algorithm. It is interesting to see from Table 3that, in four cases, the elapsed time for the SAP algorithm is less larger than that for the Landweber algorithm, and is shorter than that for the Newton-Raphson algorithm.

Table 3. Elapsed time (in seconds).

Original	Newton-Raphson	Landweber	SAP
1	10.77s	3.61s	5.09s
	500 iterations	100 iterations	80 iterations
2	11.12s	4.98s	9.10s
	500 iterations	130 iterations	120 iterations
3	10.90s	5.04s	9.24s
	500 iterations	130 iterations	120 iterations
4	14.05s	7.28s	12.55s
	800 iterations	200 iterations	150 iterations
5	12.18	33.7s	16.08s
	600 iterations	5000 iterations	180 iterations

DISCUSSION ON CHAOS SEARCH

The chaotic movement is a kind of non-periodic bounded dynamic activity sensitive to initial conditions in definite systems. The chaotic movement is featured in false randomness, ergodicity and sensitivity to boundaries, etc. (Liu, Wang, & Tan, 2006). Chaotic theory has been widely used in particle swarm Optimization algorithm (Alatas, Akin, & Ozer, 2009). When it is determined that PSO algorithm is under premature convergence conditions, the diversity of groups and sustaining searching capabilities of particles can be improved with the chaotic search.

Chaotic Annealing Particle Swarm Optimization

Many rules can cause chaos and the representative chaotic model is Logistic equation (Chen & Zhao, 2009).

$$z_d^{k+1} = u \cdot z_d^k \cdot (1 - Z_d^k) \qquad k = 0,1,2,\ldots k_{max} \tag{22}$$

where u is absorption factor, k is iteration times, and d is dimensions in question. When $u = 4, zd^0 \in (0,1)$, the equation (22) will be completely under chaos conditions and $Z_d^{k=1}$ shall tour through (0,1) along with the iteration process.

Application of the chaotic search in SAP algorithm is represented in the following two procedures:

1. Initialization of algorithm: generally, the scale of chaotic group *mCHOAS* is 0.3 times of the original group scale. We need set up the maximum iterations k_{max} and give initial values to *zdk* with minor differences separately and then get *d* chaotic variables *zdk*with different tracks.
2. Searching with chaotic variables: the chaotic variables are mapped from chaotic space to the solution space for optimization of questions and chaotic search. After each step of iteration, a solution may be created for the solution space.

With k_{max} iteration, the solution mapped from chaotic variables will tour through the solution space for optimization questions and thus make full optimization.

The mapping relationship between the variable $xd \in [xd_{,min}, xd_{,max}]$ and chaotic variable $zd \in [zd_{,min}, zd_{,max}]$ shall be specified by equation (23) and (24) (Krilc, Vesterstrom, & Riget, 2002).

$$x_d = x_{d,min} + \frac{(z_d - z_{d,min})(x_{d,max} - x_{d,min})}{z_{d,max} - z_{d,min}} \tag{23}$$

$$z_d = z_{d,min} + \frac{(x_d - x_{d,min})(z_{d,max} - z_{d,min})}{x_{d,max} - x_{d,min}} \tag{24}$$

The chaotic search is carried out when the particle swarm algorithm is under premature convergence conditions, i.e. making secondary search in small field neighboring local optimal values so as to get away from local optimal values. Premature convergence is used to search for the overall optimal values with particles. All particles are trending to be collected to the same extreme point, thus the diversity of particles is gradually decreased. If such extreme value is locally optimal one, the algorithm is under premature convergence conditions. In this paper, we take the average particle distance *D*(*t*) and the fitness value variance δ^2 (Lv & Hou, 2004) as the conditions for determination of premature convergence.

$$D(t) = \frac{1}{l \times m} \sum_{i=1}^{m} \sqrt{\sum_{d=1}^{D} (x_{id} - \overline{x}_d)^2} \tag{25}$$

where *l* is the length of diagonals of the search space, *m* is group scale, *D* is the dimension of solution space, *xid* is *dth* dimension coordinates of *ith* particle, and $\overline{x}_d$ is the average value of *xid*.

$$\delta^2 = \frac{1}{m}\sum_{i=1}^{m}\left(\frac{f_i - f_{avg}}{f}\right)^2$$

$$f = \begin{cases} \max\limits_{1 \le i \le m}\left|f_i - f_{avg}\right| & \max\limits_{1 \le i \le m}\left|f_i - f_{avg}\right| > 1 \\ 1 & other \end{cases} \tag{26}$$

where *fi* is the value of fitness function of *ith* particle, *favg* is average value of fitness functions of all particles. When δ^2 or $D(t)$ is less than the given thresholds, it may be considered that the algorithm has been under premature convergence conditions.

Performance Analysis of CAPSO Algorithm

In order to analyse the performance of chaotic search in SAP algorithm, we take SAP algorithm to make image reconstruction for typical flow patterns (core flow, bubble flow, laminar flow and circular flow), and then compared them with the imaging results of CAPSO algorithm. The imaging results are shown in Table 4. From Table 4, we can see that imaging results of the two kinds of algorithms are almost the same. It indicates that SAP algorithm embedding chaotic search did not significantly improve the imaging precision.

Table 4. Imaging results of SAP and CAPS.

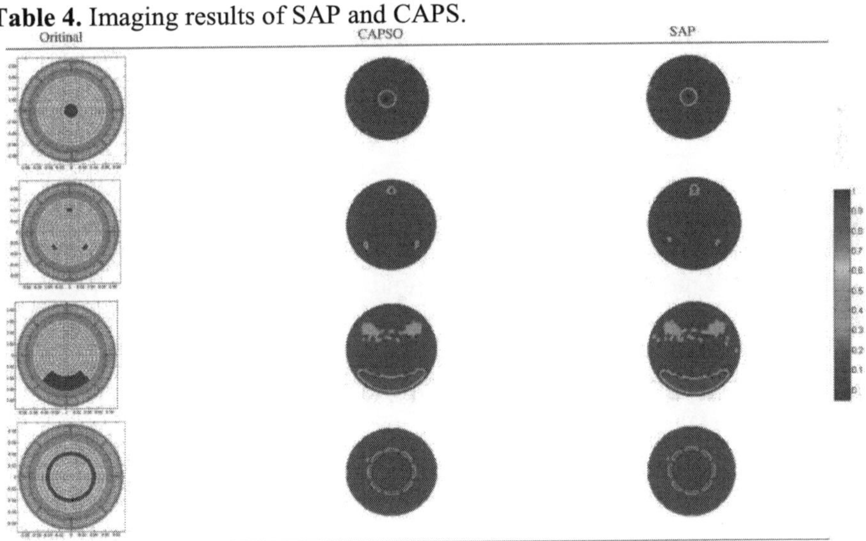

In the worst case, the time complexity of the SAP and CAPSO algorithms are$O(t_{max}$x3x$mx2D)$ and $O(t_{max}$x3x$mx2Dxm_{chaos}$xk_{max}x$D)$, where, m is group scale, D is the dimension of particles, t_{max} is the max iterations, m_{chaos} is the chaotic group scale, andk_{max} is the maximum chaotic iterations.

Obviously, the time complexity of the CAPSO algorithm is much greater than that of the SAP algorithm.

CONCLUSIONS

In this paper, we have introduced an ECT image reconstruction algorithm on basis of LS-SVM and APSO, named SAP algorithm. In order to propose this algorithm, we introduced annealing ideas into particle swarm optimization algorithm, taking cooling process function to replace inertia factor and constructing the time variant inertia weight function featured in annealing mechanism. As errors caused by the fixed sensitivity matrix for ECT reverse questions, we took LS-SVM to exercise for the errors and apply exercise results to the improvement of APSO. Then the optimized resolution of reconstructed images was searched. The experimental results demonstrated that using SAP algorithm can get high precise reconstructed images.

Finally, in order to further improved the precision of SAP algorithm, chaotic search process was merged into SAP. Experimental results demonstrated that though SAP algorithm embedding chaotic search did not significantly improve the imaging precision. However the time complexity of the algorithm is greatly improved.

ACKNOWLEDGEMENTS

This work is financially supported by Projects 51405381 and 51475013 from the National Natural Science Foundation of China, Project 201314 supported by Engagement Foundation of Xi'an University of Science and Technology, and a Project supported by Scientific Research Foundation for Returned Scholars, Ministry of Education of China 2011508

REFERENCES

1. Arce, A., Covarrubias, D.H., Panduro, M.A., & Garza, L.A. (2012). Design of beam-forming networks for multibeam antenna arrays using coherently radiating periodic structures. In Proceedings of International Congress on 1st Instrumentation and Human Applied Science, 10, 48-56.

2. Alatas, B., Akin, E., & Ozer, A.B. (2009). Chaos embedded particle swarm Optimization algorithms. Chaos Solutions & Fractals, 40, 1715-1734.
3. Chen, C., & Zhao, C.X. (2009). Path test data generation based on chaos anneal particle swarm optimization. Journal of Nanjing University of Science and Technology, 3, 376-381.
4. Eberhart, R., & Kennedy, J. (1995). A new optimizer using particles swarm theory. In Proceedings of the Sixth International Symposium on Micro Machine and Human Science, 7, 39-43.
5. Eberhart, R., & Shi, Y. (1998). Parameter selection in particle swarm optimization. Proceedings of Evolutionary Programming, 1447, 591- 600.
6. Eberhart, R., & Shi, Y. (2000). Comparing inertia weights and constriction factors in Particle Swarm Optimization. Proceedings of the Congress on Evolutionary Computation, 1, 84-88.
7. Griffiths, H. (1988). A phantom for electrical impedance tomography. Clinicalphysics and Physiological Measurement, 9, 15-20.
8. Gerardo, A., Mauricio, O.C., Nareli, C.C, Ricardo, B.F., & Jessus, A.A.C. (2009). Comparative study of parallel variants for a particle swarm optimization algorithm implemented on a multithreading GPU. Journal of Applied Research and Technology, 7, 292-309.
9. Gu, Y.P., Zhao, W.J., & Wu Z.S. (2010). Least squares vector machine algorithm. Journal of Tsinghua University (Science & Technology), 50, 1063-1066.
10. Kirkpatrick, S. (1983). Optimization by simulated annealing: quantitative studies. Journal of Statistical Physics. 34, 975-986.
11. Krilc, T., Vesterstrom, J.S., & Riget, J. (2002). Particle swarm optimization with spatial particle extension. Proceedings of the World on Congress on computational Intelligence, 1474-1479.
12. Lv, Z.X., & Hou, Z.R. (2004). Particle swarm optimization with adaptive mutation. Acta Electronic Sinica, 32, 416-420.
13. Lin, J.S. (2001). Annealed chaotic neural network with nonlinear selffeedbackand its application to clustering problem. Pattern Recognition, 34, 1093-1104.
14. Liu, H.B., Wang, X.H., & Tan, G. (2006). Convergence analysis of particle swarm optimization and its improved algorithm based on chaos. Control and Decision, 21, 636-645.
15. Ma, M., Wang, H.X., & Tian, L.M. (2006). Electrical capacitance tomography system based on DSP. Chinese Journal of Sensors and Actuators, 19, 705-708.
16. Ott, E., Grebogi, C., & Yoke, J.A. (1990). Controlling chaos. Physical Review Letters. 64, 1196-1199.
17. Peng, L.H., Lu, D., & Yang, W.Q. (2004). Image reconstruction algorithms for electrical capacitance tomography: state of the art. Journal of Tsinghua University (Science & Technology), 44, 478-484.
18. Peng, L.H., Jiang, P., Lu, G., & Xiao, D. (2007). Window functionbased regularization for electrical capacitance tomography image reconstruction. Flow Meas Instrum, 18, 277-284.

19. Rezazadeh, H., Ghazanfri, M., & Sadjadi, S.J. (2009). Linear programming embedded particle swarm optimization for solving an extended model of dynamic virtual cellular manufacturing systems. Journal of Applied Research and Technology. 7, 83-108.
20. Sait, S.M., Sheikh, A.T., & ElMaleh, A.H. (2013). Cell assignment in hybrid CMOS/Nanodevices architecture using a PSO/SA hybrid algorithm. Applied Research and Technology, 11, 653-664.
21. Tikhonov, A.N., & Arsenin, V.Y. (1977). Solutions of ill-posed problems (1st ed.). Washington, DC: Winston.
22. Wang, H.X., Zhu, X.M., & Zhang, L.F. (2005). Conjugate gradient algorithm for electrical capacitance tomography. Tianjin University, 38, 1-4.
23. York, T. (2001). Status of electrical tomography in industrial applications. Electronic Imaging, 10, 600-619.
24. Yang, W.Q. (1997). Modeling of capacitance tomography sensors. IEE Proceedings: Science, Measurement and Technology. 114, 203-208.
25. Yang, W.Q., Spink, D.M., York, T.A., & McCann, H. (1999). An image reconstruction algorithm based on Landweber's iteration method for electrical capacitance tomography. Measurement Science and Technology, 10, 1065-1069.
26. Yang, W.Q., & Peng, L.H. (2003). Image reconstruction algorithms for electrical capacitance tomography. Measurement Science and Technology, 14, R1-R13.
27. Yang, W.Q. (1996). Hardware design of electrical capacitance tomography systems. Measurement Science and Technology, 7, 225-232.
28. Yang, W.Q. (2001). Further developments in an ac-based capacitance tomography system. Review of Scientific Instruments, 72, 3902-3907.
29. Yang. D.Y., Xu, C.L., & Zhou, B. (2010). Electrical capacitance tomography system based on single measurement channel. Chinese Journal of Scientific Instrument, 31, 132-136.
30. Yeung, H., & Ibrahim, A. (2003). Multiphase flows sensor response database. Flow Measurement and Instrumentation, 14, 219-223.

CITATION

P. Wanga, J.S. Lin, M. Wang, An image reconstruction algorithm for electrical capacitance tomography based on simulated annealing particle swarm optimization, doi:10.1016/j.jart.2015.06.018.

CHAPTER 7

Mitigation of Unbalanced Voltage Sags and Voltage Unbalance in CIGRE Low Voltage Distribution Network*

Ghullam Mustafa Bhutto[1,2], *Birgitte Bak-Jensen*[1], *Pukar Mahat*[1], *Carlo Cecati*[3]

[1]Department of Energy Technology, Aalborg University, Aalborg, Denmark
[2]Department of Electrical Power, Quaid-E-Awam University of Sciences & Technology, Nawabshah, Pakistan
[3]Department of Information Engineering, Computer Sciences and Mathematics, Via G. Gronchi, L'Aquila, Italy

ABSTRACT

Any problem with voltage in a power network is undesirable as it aggravates the quality of the power. Power electronic devices such as Voltage Source Converter (VSC) based Static Synchronous Compensator (STATCOM) etc. can be used to mitigate the voltage problems in the distribution system. The voltage problems dealt with in this paper are to show how to mitigate unbalanced voltage sags and voltage unbalance in the CIGRE Low Voltage (LV) test network and networks like this. The voltage unbalances, for the tested cases in the CIGRE LV test network are mainly due to single phase loads and due to unbalanced faults. The compensation of unbalanced voltage sags and voltage unbalance in the CIGRE distribution network is done by using the four STATCOM compensators already existing in the test grid. The simulations are carried out in Dig SILENT power factory software version 15.0.

INTRODUCTION

Power quality (PQ) is one of the primary objectives of the modern power system. Among the various phenomena responsible of performance degradation, Voltage Sags (VS) defined by IEEE standards 1159-1995 as the reduction in the value of RMS voltage between 0.1 to 0.9 p.u at the power frequency with the duration from 0.5 cycles to 1 min represent one of the most important causes of poor power quality [1]. The main causes of voltage sags are short circuit faults occurring in transmission or distribution systems, transformer energizing, switching of the capacitor banks and starting of large induction motors [2]. Their effects are multiple and change with different loads. The case of sensitive house hold loads has been deeply described in [3].

Voltage sags can be symmetric or asymmetric depending on the type of short circuit faults. The majority of faults in power system are single-phase-to-ground faults [4], and consequently result in unbalanced sags. A symmetric fault involves all kinds of sequence quantities: positive, negative and zero sequence, whereas voltage sags due to symmetric fault contain only positive sequence quantities. Depending on both the type of faults and the transformer connections between medium voltage and the low voltage, different types of voltage sags can be distinguished. A detailed discussion about the influence of the transformer winding connection on the propagation of voltage sags is presented in [5, 6]. The transformer winding (delta/wye) connection used in this network blocks the flow of zero sequence components from the voltage [5]. An extended analysis of voltage sag and their classification has been carried out in [7-9].

Voltage in power system can be unbalanced due to the several reasons. One of the major reasons of voltage unbalance is an uneven distribution of single phase loads that draw unbalanced currents from the system [10, 11]. These unbalance currents will create unequal heating in each of the phases which creates unbalance heating in cables and other parts of the network, which might reduce the life time of the cables and other components [12,13]. Another reason for voltage unbalance is due to unbalanced faults. The unbalance in the voltage due to this reason is severe.

According to European standards, the percent Voltage Unbalance Factor (VUF) is defined by the ratio of the negative sequence voltage to the positive sequence voltage [14-16].

$$\%VUF = V_2/V_1 * 100 \tag{1}$$

where V_1 and V_2 are the positive and negative sequence components of the voltage respectively. According to IEEE Std. 1547.2 - 2008 the voltage unbalance factor should be below “2% to 3%”.

The aim of this paper is to investigate how to mitigate unbalanced voltage sags and also voltage unbalance in a CIGRE low voltage distribution network. This is done by using custom power devices: two D-STATCOMs and two BESS-STATCOMs used at different locations of the network.

The paper is organized as follows: Section 2 gives the description of CIGRE LV distribution network, section 3 presents the description about the control structure; Section 4 presents the simulation results without using controllers. Mitigation of the voltage sag and the voltage unbalance for the two cases (i.e. when voltage sag of 42.3% and 43.4% on phase A and phase B of bus R1 and when voltage sag of 13% on phase B of the same bus) is proposed by using appropriate controllers in section 5. Finally, the conclusion about the paper is presented in section 6.

DESCRIPTION OF CIGRE NETWORK

In order to achieve the described goal, a test distribution network set up by CIGRE comprising Wind Turbine Generator (WTG), Photovoltaic (PV) solar generation units, two batteries and STATCOMs comprising Voltage Source Converters (VSCs) at different locations has been chosen for the study [17]. Unbalanced loads are aggregated at the 0.4 kV voltage levels and are connected at bus RC, RD, R11, R15 and bus R17. The detailed data concerning bus bars, cables/lines and loads is given in [17]. The Distributed Generation (DG) units are integrated into the grid using a delta-wye with neutral (D-YN) transformer used at the beginning of the radial as shown in Figure 1. The neutral of the transformer is grounded with low impedance, $Z = (0.0032 + j0.0128)\Omega$. The single line diagram of this distribution system is shown in Figure 1. The considered network is details are described in [17].

In case of a WTG connected to an unbalanced voltage, the stator currents will be unbalanced. These unbalanced stator currents creates unequal heating in stator winding which might degrade winding insulation and thereby reducing the life time of the stator winding. The impacts of voltage unbalance on WTG's have been studied in [18]. The impacts of voltage unbalance on PV inverters are studied in [19, 20]. The CIGRE network is modeled in DIgSILENT power factory software version 15.0.

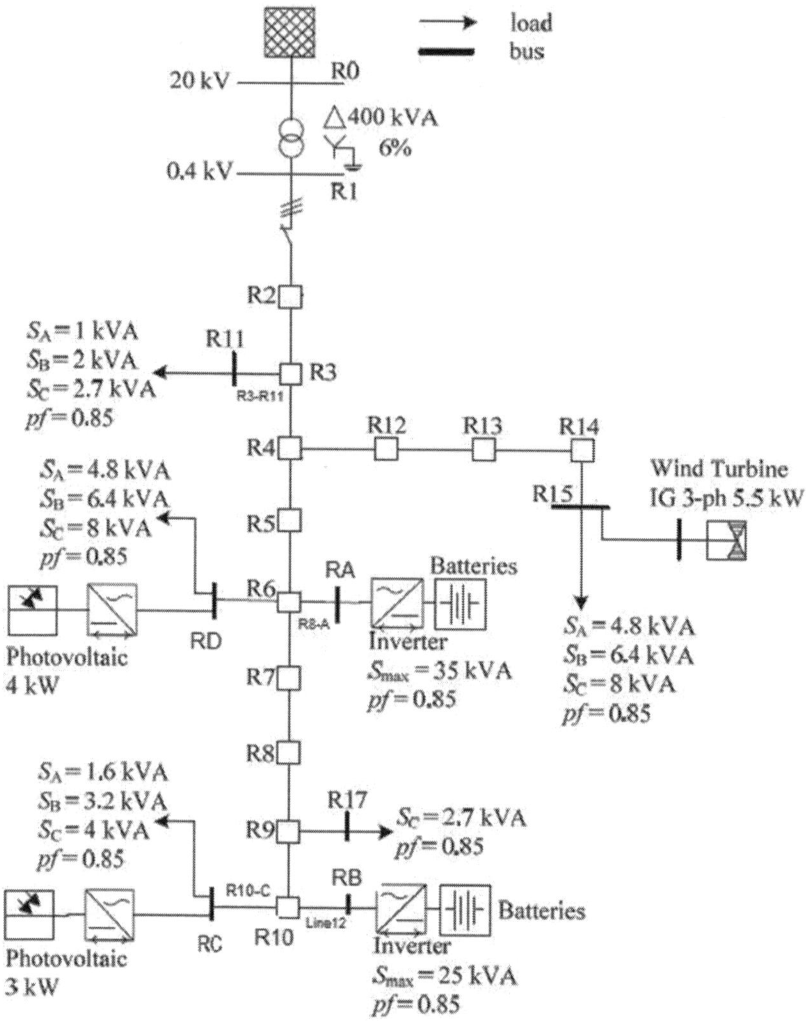

Figure 1. Single line diagram of CIGRE LV distribution system.

DESCRIPTION OF CONTROL SYSTEM

The control system developed by the author used for this study is detailed described [21]. A brief introduction of the control system is presented in this paper only. The D-STATCOM controllers for both PV systems have been developed in order to control dc-link and AC voltages by injecting/absorbing active and reactive powers respecttively [21]. The controllers for the Battery Energy Storage Systems (BESS) are developed and are able to charge/discharge the batteries at different charging rates. The BESS controllers are developed in such a way that they counteract voltage and frequency disturbances by receiving/delivering active and reactive powers [21]. The inverters of the PV units and battery units are modeled in such a way that they can deliver unbalance currents in order to mitigate unbalanced voltage sags and prevent the problems of the voltage unbalance in the network [21]. The focus in this paper is only on the control of the AC side of the voltage. The problems of the dc-link voltages on the PV inverters are discussed in [19,21] and charging/ discharging of the batteries are described in [21].

STUDY OF THE SYSTEM WITHOUT USING CONTROLLERS

To have a base study case, all the controllers have been disabled in order to study the impacts of unbalanced loads and unbalanced faults on the voltage in different parts of the low voltage CIGRE network. The voltage unbalance exists from the beginning as unbalanced loads are connected to the network. Next, a single line to ground fault on phase A with a 0 Ω fault impedance is applied at time equal to t = 2 s on the 20 kV Medium Voltage (i.e. R0) bus in order to simulate a case of severe unbalance. The fault is cleared at 2.15 s. The magnitude and phase angle of voltage in the three phases at the bus R0 is shown in Figures 2(a) and (b).

It can be seen that voltage in all three phases of the bus R0 are nearly the same during normal operating conditions as there is only minor load unbalance. At time equal to t = 2 s, the voltage in the effected phase becomes zero and the voltage in the other two phases remain nearly the same as shown in Figure 2(a). The results shown in Figure 2(a) are matching with the reference [5, 22].

The voltage angles in its three phases on bus R0 are shown in Figure 2(b). It can be seen inFigure 2(b) that the angle of the voltage is changed in the affected phase only, whereas the voltage angles of the healthy phases remains unchanged.

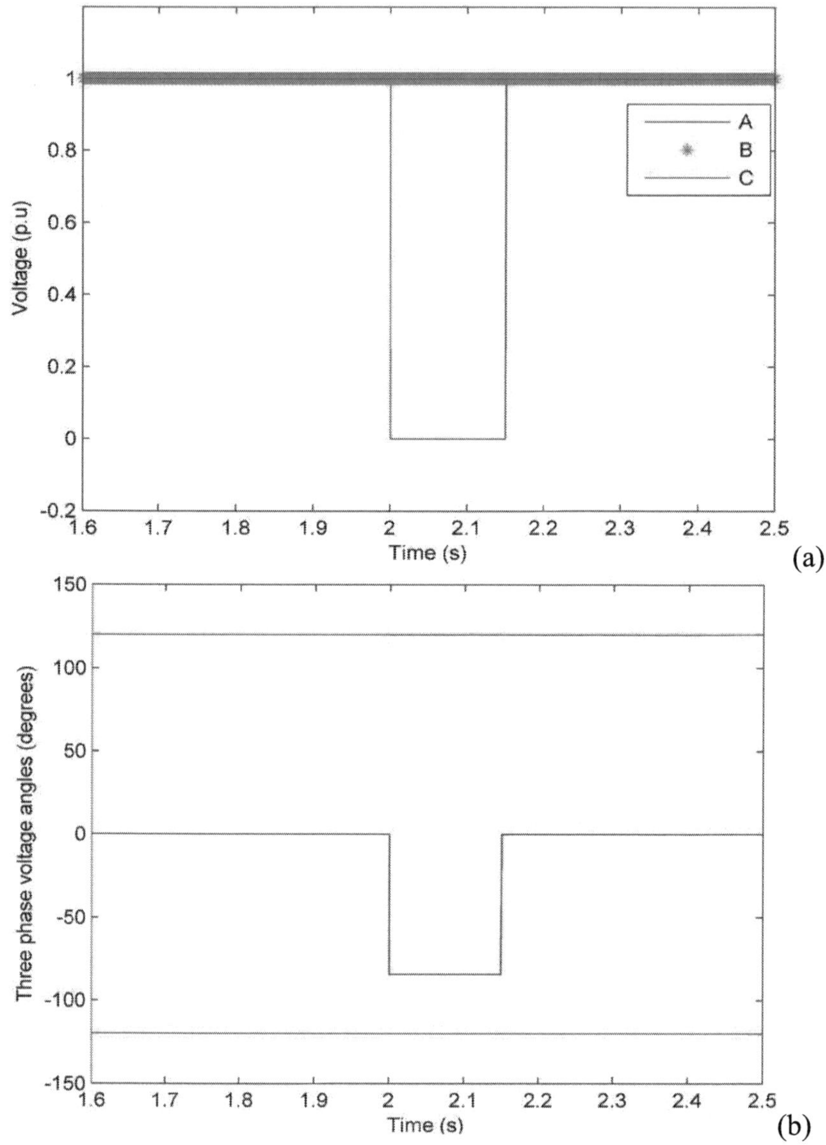

Figure 2. (a) Voltage in the three phases of the R0 bus for single phase to ground fault on phase A

with $z_f = 0$ Ω. (b) Angles of three phase voltages on bus R0 in case L-G fault on phase A with $z_f = 0$ Ω (blue line: phase A, green line: phase B, red line: phase C).

The transformation of voltages from primary (i.e. delta) of the transformer to its secondary (i.e. star) is described with the help of figure 3 [23].

In delta wye configuration of the transformer, the value of phase to neutral voltage on secondary (i.e. shown with colored lines R, Y and B) is equal to that of the line voltage at the primary divided by $\sqrt{3}$ [24]. The line to line primary voltages are given in "Equation (2)".

$$
\begin{aligned}
V_{ba} &= V_b - V_a \\
V_{cb} &= V_c - V_b \\
V_{ac} &= V_a - V_c
\end{aligned} \qquad (2)
$$

Putting the values of phase voltages with respect to their phase angles obtained from Figure 2 in "2," line to line voltages on the primary of the transformer can be calculated.

$$
\begin{aligned}
V_{ba} &= 1 < -120^0 - 0 < -84^0 \\
V_{ba} &= 1 < -120^0
\end{aligned}
$$

The voltage V_{ba} on the primary corresponds to the voltage in phase B on the secondary of the transformer as shown in Figure 3 and is given by $V_{bN} = V_{ba}/\sqrt{3}$ which is equal to $0.578 < -120^0$ p.u.

$$
\begin{aligned}
V_{cb} &= 0.999 < 120^0 - 1 < -120^0 \\
V_{cb} &= 1.723 < 90^0
\end{aligned}
$$

The voltage V_{cb} on the primary corresponds to the voltage in phase C on secondary of the transformer as shown in Figure 3 and is given by $V_{cN} = V_{cb}/\sqrt{3}$ which is equal to $0.999 < 90^0$ p.u. $V_{ac} = 0 - 0.999 < 120^0$ $V_{ac} = 0.999 < -60^0$

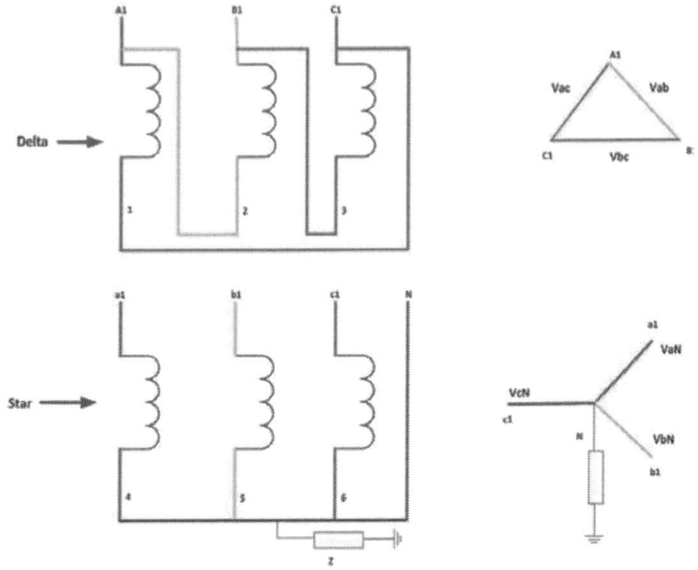

Figure 3. The transformation of voltages from primary of DYN1 transformer to secondary.

The voltage V_{ac} on the primary corresponds to the voltage in phase A on the secondary of the transformer as shown in Figure 3 and is given by $V_{aN} = V_{ac}/\sqrt{3}$ which is equal to $0.578 < -60^0$ p.u.

The voltage in the three phases at the bus R1 is shown in Figure 4. At t = 2 s, the voltage profile on this bus changes as shown in Figure 4.

It can be seen in Figure 4 that there is voltage sag of 42.3% in phase A and 42.4% in phase B on the bus R1. Phase C on this bus is unaffected. A slight difference in voltage sag on phase A and B is due to the slight difference in their loadings. The load on phase B is slight more than the load on phase A, so voltage sag on phase B is slightly deeper than on phase A. When the fault is cleared after 150 m s, the voltage in all the phases return to prefault values. The results shown in figure 4 also match with [5,6,9].

The voltage at different phases on bus RA, RB, RC and Bus RD is shown in Figures 5(a)-(d) respectively. Similar to Figure 4, there is mild unbalance in voltage in normal operating conditions due to unbalanced loads and severe unbalance due to the unbalanced fault.

Table 1 shows the unbalance factor in different parts of CIGRE distribution network under mild and severe unbalance.
It can be seen in Table 1 that there is a mild unbalance due to the unbalanced loads connected on different buses in normal operating condition and severe unbalance in the network grid due to a fault. The difference in the unbalance factor during normal operating conditions is due to different loadings on the buses in the network grid.

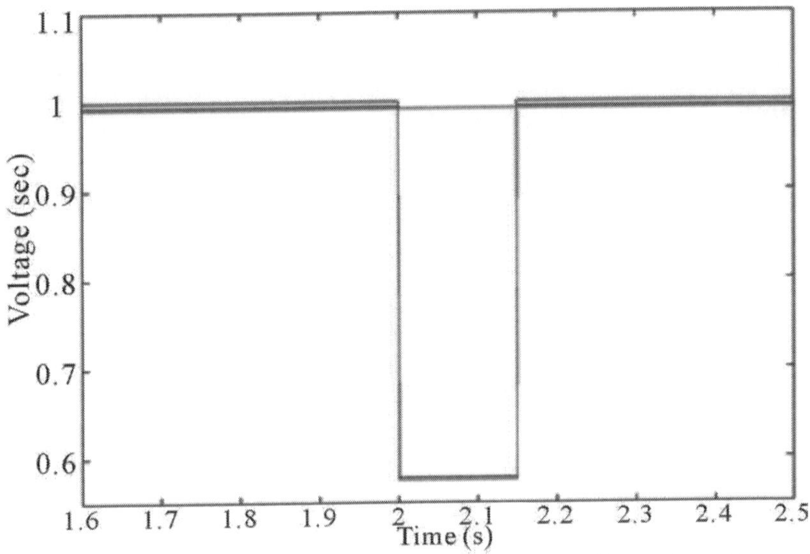

Figure 4. Voltage on R1 bus (blue line: phase A, green line: phase B, red line: phase C).

Table 1. Unbalance Factors in different parts of network for the two conditions.

Bus No	%(VUF) in normal operating conditions	%(VUF) during fault
RA	0.45	49.64
RB	0.48	49.6
RC	0.5	49.59
RD	0.43	49.638
R11	0.31	49.66
R15	0.395	49.74
R17	0.55	49.55

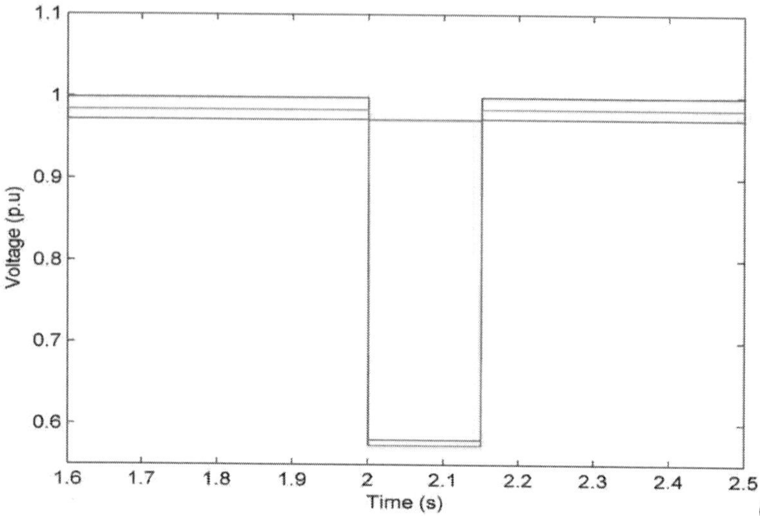
Voltage (p.u)
1.1
1
0.9
0.8
0.7
0.6
1.6
1.7
1.8
1.9
2
2.1
2.2
2.3
2.4
2.5
Time (s)
(a)

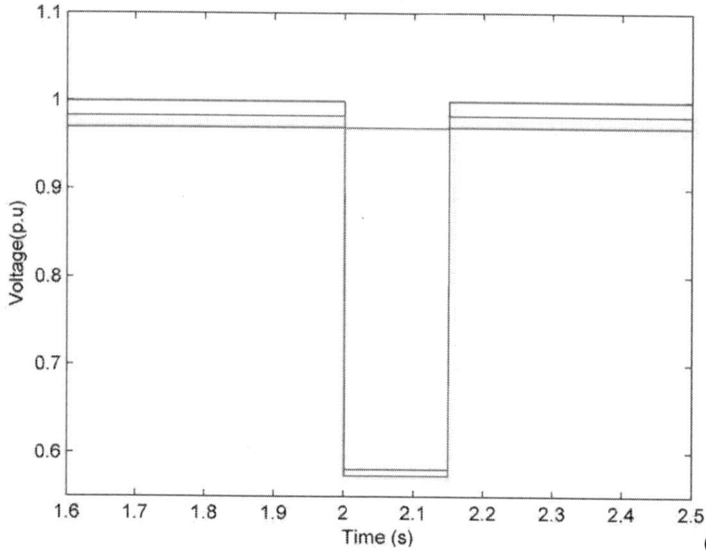
Voltage(p.u)
1.1
1
0.9
0.8
0.7
0.6
1.6
1.7
1.8
1.9
2
2.1
2.2
2.3
2.4
2.5
Time (s)
(b)

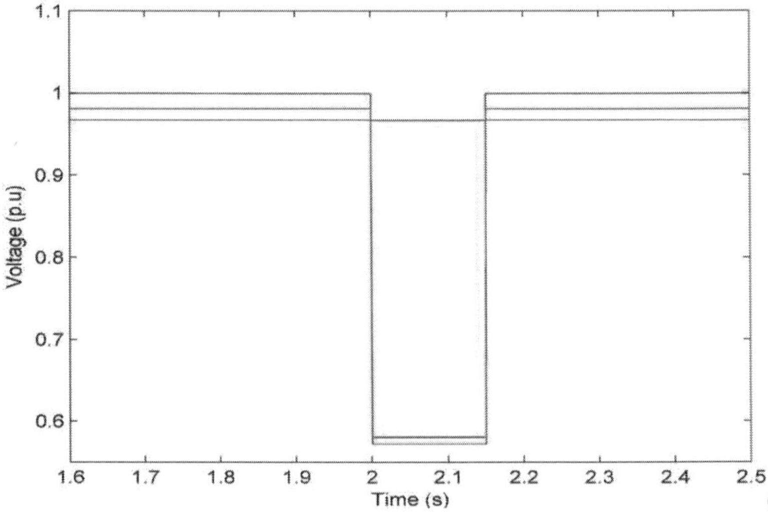

(c)

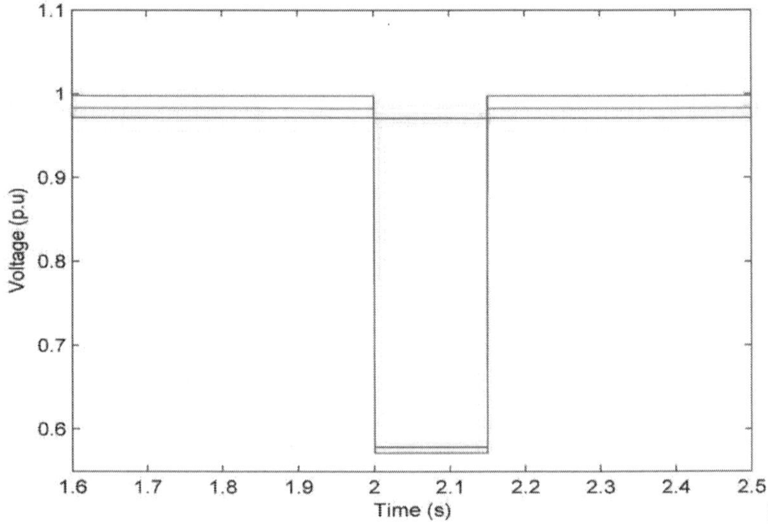

(d)

Figure 5. (a) Voltage on bus RA. (b) Voltage on bus RB. (c) Voltage on bus RC. (d) Voltage on bus RD. (blue line: phase A, green line: phase B, red line: phase C).

THE MITIGATION OF UNBALANCE VOLTAGE SAG AND VOLTAGE UNBALANCE

Four STATCOM controllers are used in order to inject/absorb the desired amount of reactive power in each of the phases in order to restore the voltage of different phases to permissible limits and mitigate the problems of voltage unbalance. The reactive power injected/absorbed by all of the four VSCs in the three phases of bus RA, RB, RC and bus RD are shown in Figures 6(a)-(d) respectively.

Since all four inverters are modeled for voltage regulation, they inject or absorb reactive power in different phases of the buses where they are connected. If the voltage of the bus in any of the phase is greater than the voltage of an inverter connected on that bus, the reactive power in that phase is absorbed by the inverter and vice versa [25]. Two of the inverters (i.e. VSC1 and VSC2) are used for battery applications and the other two (VSC3 and VSC4) for the PV units.

The contribution of reactive power from the different converters (i.e. VSC1, VSC2, VSC3 and VSC4) is according to their rated power. The power and voltage ratings of the inverters used in the network are shown in Table. 2. The inverters mentioned in Table 2 can deliver reactive power up to their rated value if they do not inject or absorb active power. The batteries are charged at very slower rates (i.e. 1/100 rate) so that battery inverters can provide maximum amount of reactive power in order to mitigate voltage sags. The PV inverters are also set to deliver maximum amount of reactive power.

It can be seen in Figure 6 that reactive power is injected in different phases of the bus RA, RB, RC and bus RD in the normal operating conditions in order to reduce the unbalance factor. As shown in Figures 6(a) and (b), VSC1 and VSC2 deliver small amounts of reactive power during steady state condition but inject/absorb maximum amount of reactive power during unbalanced voltage sag. Since the two phases (i.e. phase A and phase B) have seen voltage sag in the network buses therefore all of the inverters are injecting reactive power in these two phases at t = 2 s as shown inFigure 6.

Table 2. The ratings of the inverters used in the network

Name of an inverter	Apparent power, S (kVA)	V_{DC} (kV)	V_{AC} (kV)
VSC1	35	0.714	0.4
VSC2	25	0.781	0.4
VSC3	18	1	0.4
VSC4	24	1	0.4

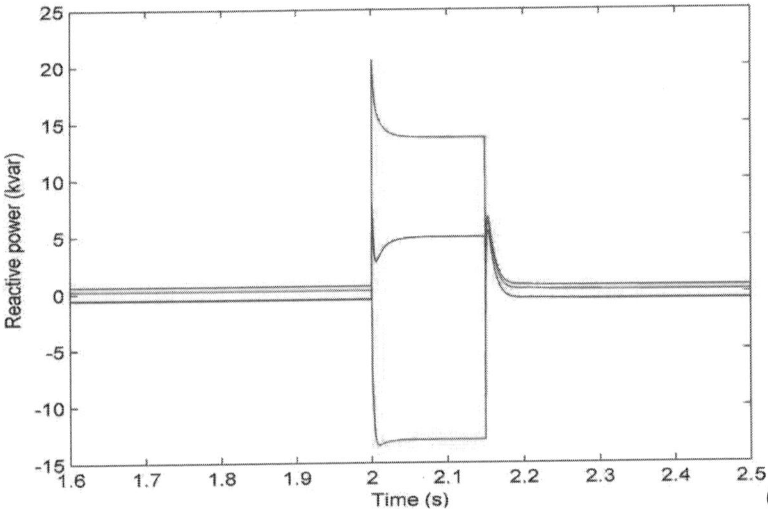
Reactive power (kvar)
25
20
15
10
5
0
-5
-10
-15
1.6
1.7
1.8
1.9
2
2.1
2.2
2.3
2.4
2.5
Time (s)

(a)

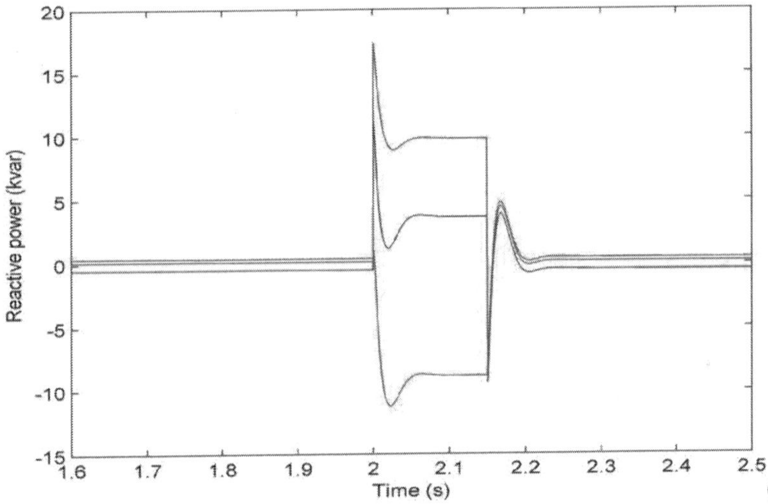
Reactive power (kvar)
20
15
10
5
0
-5
-10
-15
1.6
1.7
1.8
1.9
2
2.1
2.2
2.3
2.4
2.5
Time (s)

(b)

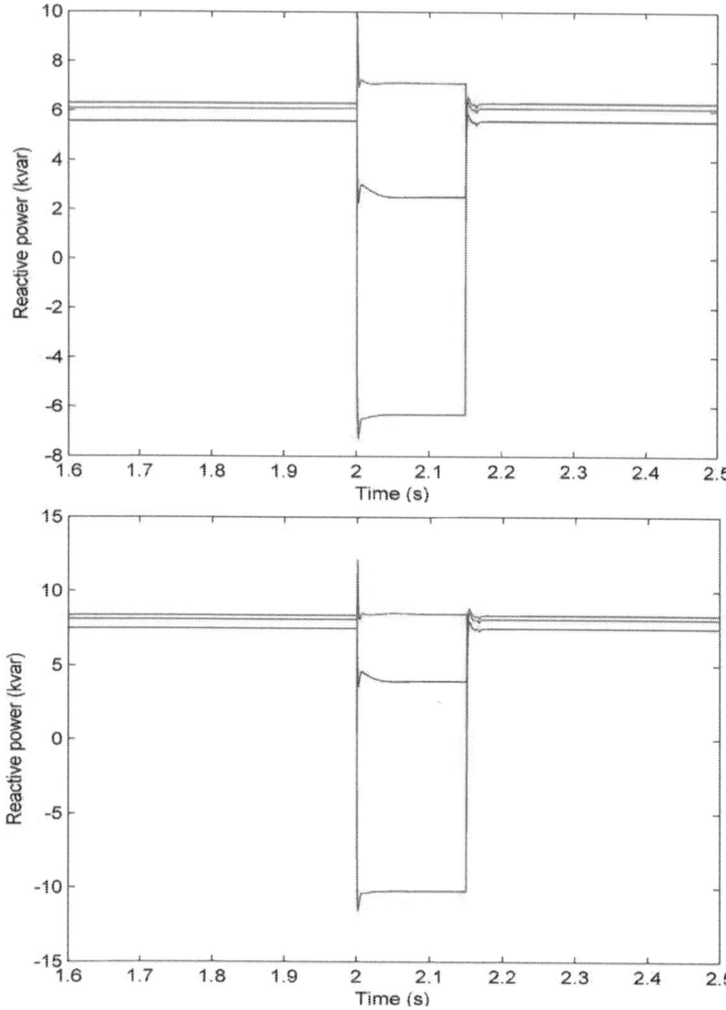

(c)

(d)

Figure 6. (a) Reactive power of VSC1 in three phases. (b) Reactive power of VSC2 in three phases. (c) Reactive power of VSC3 in three phases. (d) Reactive power of VSC4 in three phases. (blue line: phase A, green line: phase B, red line: phase C).

The reactive power injected by all of the inverters in phase B is little bit higher than phase A because voltage sag in phase B is little deeper than phase A in all branches. The phase C in all branches has not seen volt age sag so in general there should be no injection of reactive power by the inverters in this phase. But the purpose of the controllers developed for this study is not only to mitigate voltage sag but also to compensate against voltage unbalance in the network. Due to this reason the reactive power is absorbed in phase C in order to reduce unbalance factor.

When fault appears in some part of the network, there is an increase in the flow of the currents delivered by the inverters used in the network. The rise of the current in the inverters is controlled by using current limiters so the IGBTs may not damage against over currents. The peaks in the reactive powers of VSC1 and VSC2 in each of the phase as shown in Figures 6(a) and (b) are due to this phenomenon.

It can be seen in Figures 6(c) and (d) that VSC3 and VSC4 are injecting nearly full amount of reactive power in steady state operation in order to meet the line voltage drops and other load unbalances in the network. When the unbalanced voltage sag appears at t = 2 s, the controllers do not have enough reactive power to compensate for it and hence the distribution system operates in under voltage conditions across these points. Due to this reason the reactive power injected by these controllers in some of the phases decrease further (i.e. $Q = VI \sin\phi$) as seen in Figures 6(c) and (d). The peaks in Figures 6(c) and (d) at the beginning and at the end of voltage sag are due to the charging and discharging of DC-link capacitors in order to maintain DC-link voltage of PV1 and PV2.

The new value of voltage in each of the phases in per unit in different parts on network after using compensation and the voltage unbalance factors in normal and fault periods are shown in Table 3.

Table 3 shows that the controllers to some extent have mitigated the voltage sags in the different phases and have improved the unbalance factors. The unbalance factors in fault conditions are still not within an acceptable range and also the voltages in phase A and phase B are not improved up to the acceptable limits [26] because the converters do not have enough reactive power capacity to mitigate voltage sag and voltage unbalances of such big depth.

To figure out the limit of the converters in relation to the voltage sag depth after voltage sag compensation, a single phase to ground fault on phase A at the R0 bus with a fault impedance of 3 Ω is applied on time equal to t = 2 s. The fault is cleared at 2.15 s. The voltages in the three phases of the R0 bus in this case are shown in Figure 7. It can be seen in Figure 7 that voltage in the effected phase during the fault is reduced but not equal to zero as in the last case.

The fault on R0 bus creates unbalanced voltage sags in different parts of the network gird. The transformation of voltage sag from the delta side of the transformer to its wye side is according toFigure 3. The voltages in three phases of bus R1 obtained by using above transformation method in this new case is shown in figure 8. The values of the voltage during fault in each of the phases have been verified by using (2) and the results are matching with the results shown in Figure 8.

For simplicity the voltage on one of the buses (i.e. bus RA) with and without controllers in this new case is shown in Figures 9(a) and (b). It can be seen in Figure 9(a) that without using controllers phase B of bus RA of the network is less than 90% which is not within a tolerable limit according to IEEE standard 1159-1995 and Danish standards [26]. By using controllers the voltages in all of the phases are restored within the permissible limits of ±10% according to Danish standards.

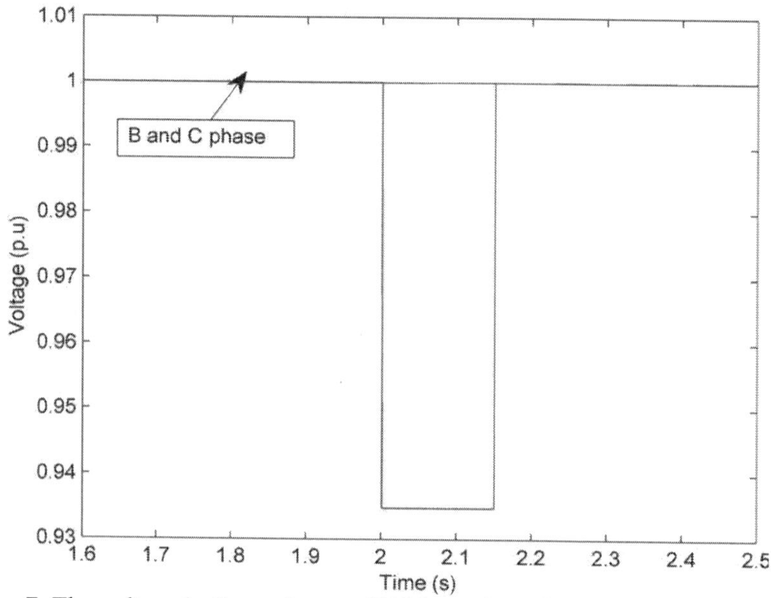

Figure 7. The voltage in three phases of R0 bus when single phase to ground fault on phase A with $Z_f = 3$ Ω. (blue line: phase A, green line: phase B, red line: phase C).

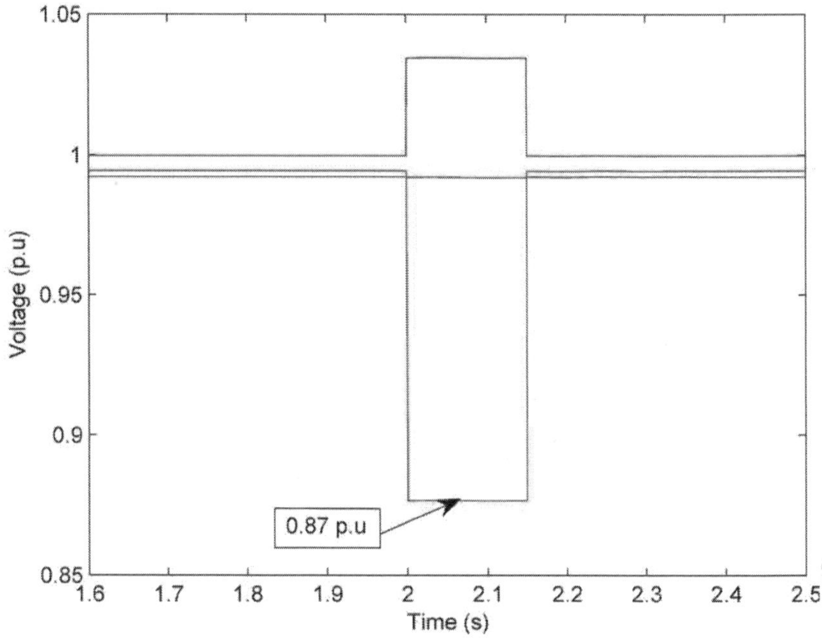

Figure 8. Voltage on the R1 bus. (blue line: phase A, green line: phase B, red line: phase C).

Table 3. Voltage in different phases in different parts of the network and unbalance factors.

Bus number	Voltage in normal conditions			%UF in normal operation	Voltage during faults			%UF during fault
	A	B	C		A	B	C	
RA	1.007	0.9986	0.9949	0.299	0.69	0.54	0.912	33.2
RB	1.0066	0.9991	0.9956	0.32	0.691	0.545	0.9087	32.4
RC	1.0078	0.9982	0.994	0.34	0.68	0.55	0.91	32.55
RD	1.0061	0.9977	0.9936	0.299	0.654	0.56	0.926	35.5
R11	1.0045	0.9976	0.9945	0.22	0.61	0.579	0.95	40.2
R15	1.0052	0.996	0.9904	0.295	0.626	0.57	0.939	38.3
R17	1.0077	0.9992	0.992	0.39	0.67	0.56	0.914	33.33

Table 4. Voltages in different phases in different parts of the network with and without controllers and unbalance.

Bus number	Voltage in normal conditions			%UF in normal operation	Voltage during faults			%UF during fault
	A	B	C		A	B	C	
RA	1.034	0.867	0.97	9.22	1.0366	0.919	0.97	6.46
RB	1.034	0.866	0.969	9.2	1.0367	0.92	0.97	6.3
RC	1.034	0.864	0.966	9.26	1.037	0.918	0.97	6.38
RD	1.033	0.866	0.97	9.4	1.0356	0.91	0.979	6.9
R11	1.034	0.87	0.98	9.27	1.033	0.90	0.988	7.69
R15	1.0336	0.879	0.974	9.45	1.034	0.90	0.98	7.38
R17	1.035	0.867	0.967	9.24	1.036	0.916	0.975	6.5

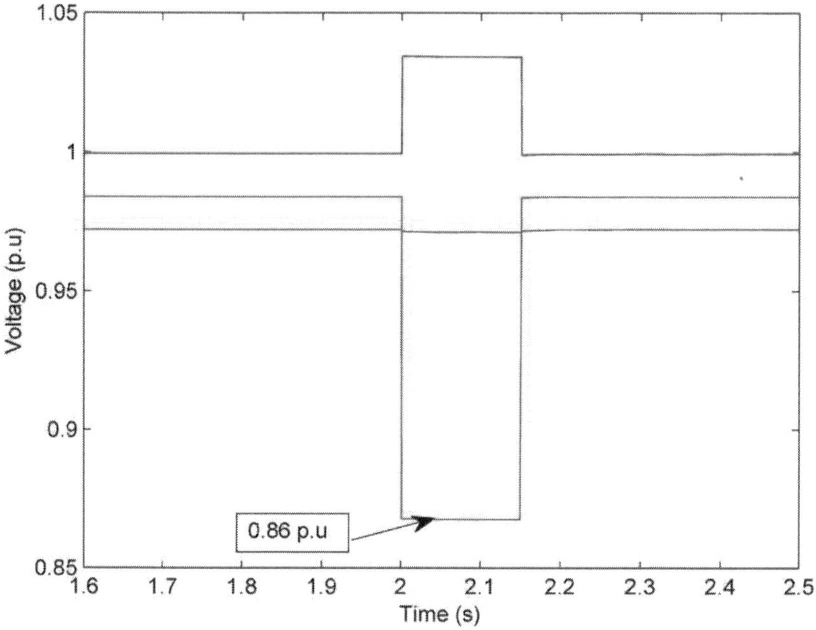

(a)

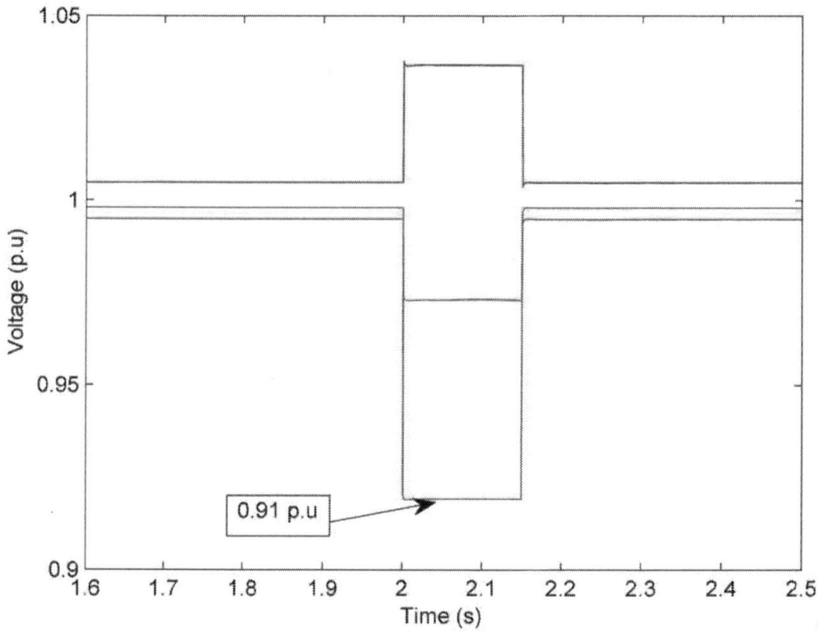

(b)

Figure 9. (a) Voltage on bus RA without controllers. (b) Voltage on bus RA with controllers. (blue line: phase A, green line: phase B, red line: phase C).

It can be seen in Table 4 that voltage in all parts of network in each of the phases are restored within permissible limits and the voltage unbalance during fault are also reduced to a smaller value by using the developed controllers. The controllers have successfully mitigated the unbalanced voltage sags in all parts of the network grid but they are unable to maintain unbalance factor according to IEEE standards 1547.2-2008 during a fault. The only way to restore the unbalance factor according to the standards is to use bigger size of the battery units rather than the existing ones.

CONCLUSION

The mitigation of the voltage unbalance during normal and faulted conditions and the mitigation of unbalanced voltage sag have been performed by using two D-STATCOM and two BESS based STATCOM controllers. The compensation for two different cases (i.e. voltage sag of 42.3% and 43.4% on phase A and phase B respectively and when voltage sag of 13% on only phase B) has been performed. It has been observed that the existing controllers can mitigate the voltage sag of up to 13% depth in any of the phases. If the depth of the voltage sag in any phase is more than that, the controllers cannot restore the voltage up to the operating limits. By using the existing controllers, the unbalance in voltage has been reduced to a small value for the case when the voltage sag in one of phases is not more than 13%, but this voltage unbalance is still not in the acceptable limits. This has been verified in simulation results. The compensation of voltage unbalance in the network grid is proposed by using bigger energy storage units rather than existing one. In future work, the controllers will be tested in island conditions as well.

ACKNOWLEDGEMENTS

We are thankful to Kai Strunz for providing the CIGRE test network for LV distribution network. I, Ghullam Mustafa am thankful to Quaid-E-Awam University of Engineering Sciences and Technology, Nawabshah, Sindh, Pakistan for the support and funding.

REFERENCES

1. IEEE Standard 1159-1995, "IEEE Recommended Practice for Monitoring Electric Power Quality," IEEE, 1995.
2. H. J. B. Math, I. Y. H. Gu, P. G. V. Axelberg and E.Styvaktakis, "Classification of Underlying Causes of Power Quality Disturbances: Deterministic versus Statistical Methods," EURASIP Journal on Applied Signal Processing, Vol. 1, 2007, pp. 172-172.
3. G. G. Karady, S. Saksena and B. Z. Shi, "Effects of Voltage Sags on House Hold Loads," IEEE Power Engineering Society General Meeting, Vol. 3, 2005, pp. 2456- 2461.

4. M. Mc Granaghan, D. R. Mueller and M. Samotyj, "Voltage Sag in Industrial Systems," IEEE Transactions on Industry Applications, Vol. 29, 1993, pp. 397-403.http://dx.doi.org/10.1109/28.216550
5. M. T. Aung and J. V. Milanovic, "The Influence of Transformer Winding Connections on the Propagation of Voltage Sags," IEEE Transactions on Power Delivery, Vol. 21, 2006.
6. W. R. Mendes, M. I. Samesima and F. A. Moura, "Influence of Power Transformer Winding Connections on the Propagation of Voltage Sags through Electric System," Proceedings of the IEEE, 2008, pp. 1-6.
7. L. D. Zhang and M. H. J. Bollen, "A Method for Characterizing Unbalanced Voltage Dips (Sags) with Symmetrical Components," IEEE Power Engineering Review, Vol. 18, No. 7, 1998, pp. 50-52. http://dx.doi.org/10.1109/MPER.1998.686958
8. C. Di Perna, P. Verde, A. Sannino, M. H. J. Bollen, "Static Series Compensator for Voltage Dip Mitigation with Zero-Sequence Injection Capability," 2003 IEEE Bologna Power Tech Conference Proceedings, Bologna, 23-26 June 2003.
9. R.Teodorescu, M. Liserre and P. Rodríguez, "Grid Synchronization in Three Phase Power Converters," In: Grid Converters for Photovoltaic and Wind Power Systems, John Wily & Sons Ltd., Chichester, 2011, pp. 169-181.
10. von Jouanne and B. B. Banerjee, "Voltage Unbalance: Power Quality Issues, Related Standards and Mitigation Techniques," Electric Power Research Institute, Palo Alto, EPRI Final Rep, 2000.
11. Aneel, "Contribution for the Normalization of the Quality of Electric Power-Harmonic and Unbalances in the Electric Networks," Ph.D. Thesis, Federal University of Uberlandia, 2000.
12. von Jouanne and B. Banerjee, "Assessment of Voltage Unbalance," IEEE Transactions on Power Delivery, Vol. 16, No. 4, 2001, pp. 782-790.
13. F. C. Pereira, J. C. de Oliveira, O. C. N. Souto, A. L.A. Vilaca and P. F. Ribeiro, "An Analysis of Costs Related to the Loss of Power Quality," Proceedings on Harmonics and Power Quality, Vol. 2, 1998, pp. 14-16.
14. V. J. Gosbell, S. Perera and V. Smith, "Voltage Unbalance," Technical Power Quality Center, University of Wollongong, Technical Note No. 6, 2002.
15. C.-Y. Lee, "Effects of Unbalanced Voltage on the Operation Performance of a Three-Phase Induction Motor," IEEE Transactions on Energy Conversion, Vol. 14, No. 2, 1999, pp. 202-208. http://dx.doi.org/10.1109/61.956770
16. I. Maswood, G. Joos, P. D. Ziogas and J. F. Lindsey, "Problems and Solutions Associated with the Operation of Phase-Controlled Rectifiers under Unbalanced Input Voltage Conditions," IEEE Transactions on Industry Applications, Vol. 27, 1991, pp. 765-772.http://dx.doi.org/10.1109/28.85494
17. Benchmark Systems for Network Integration of Renewable Energy Resources, Version 7, CIGRE Task Force C6.04.02, 2011.
18. E. Muljadi, C. P. Butterfield, T. Batan and D. Yildirim, "Understanding the Unbalanced Voltage Problems in Wind Turbine Generation," National Renewable Energy Laboratory (NREL), 1999.

19. J. Alonso-Martínez, J. Eloy-García and S. Arnaltes, “Control of a Three-Phase Grid Connected Inverter for Photovoltaic Applications with a Fuzzy MPPT under Unbalanced Conditions,” Proceedings on Power Electronics and Applications, 2009, pp. 1-7.
20. L. Fan, Z. Miao, and A. Domijan, “Impact of unbalanced grid conditions on PV systems,” IEEE power and energy society general meeting, 2010, pp. 1-6.
21. G. M. Bhutto, B. Bak-Jensen and P. Mahat, “Modeling of the CIGRE Low Voltage Test Distribution Network and the Development of Appropriate Controllers,” Transactions on Smart Grid and Clean Energy, Vol. 2, 2013, pp. 184-191.
22. D. Das, “Electrical Power System,” New Age International (P) Publisher, New Delhi, 2006.
23. Vector Group of Transformer, Electrical Notes & Articles. http://electricalnotes.wordpress.com/2012/05/23/vector-group-of-transformer.
24. W. H. Kersting, “Distribution System Modeling and Analysis”, CRC Press, New Mexico, 2002.
25. C. Hochgraf, R. H. Lasseter, “STATCOM Control for Operation with Unbalanced Voltages,” IEEE Transactions on Power Delivery, Vol. 13, 1998, pp. 538-544.http://dx.doi.org/10.1109/61.660926
26. Rekommandation 16, Spændingskvalitet I lavspændingsnet (voltage quality in low voltage Networks), 4. Udgave, 2011

CITATION

G. Bhutto, B. Bak-Jensen, P. Mahat and C. Cecati, "Mitigation of Unbalanced Voltage Sags and Voltage Unbalance in CIGRE Low Voltage Distribution Network," Energy and Power Engineering, Vol. 5 No. 9, 2013, pp. 551-559. doi: 10.4236/epe.2013.59060.

CHAPTER 8

Comparative Study of Conventional, Fuzzy Logic and Neural PID Speed Controllers with Torque Ripple Minimization for an Axial Magnetic Flux Switched Reluctance Motor

Eric S. Sanches, José A. Santisteban

Electrical Engineering Department, Mechanical Engineering Postgraduate Program (PGMEC), Fluminense Federal University, Niterói, Brazil

ABSTRACT

Three speed controllers for an axial magnetic flux switched reluctance motor with only one stator, are described and experimentally tested. As it is known, when current pulses are imposed in their windings, high ripple torque is obtained. In order to reduce this ripple, a control strategy with modified current shapes is proposed. A workbench consisting of a machine prototype and the control system based on a microcontroller was built. These controllers were: a conventional PID, a fuzzy logic PID and a neural PID type. From experimental results, the effective reduction of the torque ripple was confirmed and the performance of the controllers was compared.

INTRODUCTION

The radial magnetic flux version of switched reluctance motors (SRM) has been widely used in many variable speed industrial applications and some advantages have been reported: high torque output, wide range of operating speed, geometrical simplicity, reliability and robustness [1] [2]. In the case of an axial flux SRM (AFSRM) the air gap magnetic flux is parallel to the rotation axis. As the AFSRM can have a smaller axis length than the radial magnetic flux motor, it is believed that it is a good solution for use in applications where size is important, as in electric cars, for instance.

The switched reluctance motor (SRM) analyses are complex because of their doubly salient pole structure and nonlinear magnetic characteristics. The developed torque is a nonlinear function of the currents applied to the stator windings and their inductances, which depend on the rotor position. Nevertheless, with an appropriate control system, minimum ripple torque can be obtained.

From literature, it can be observed that many works are related to radial flux SRM; however few works are related to axial flux SRM [3]. Some of them refer to axial flux SRM with two stators [4]-[6], but in this paper the motor has only one stator. With respect to the control strategies, several strategies are reported [7]-[9].

For ordinary radial flux SRM, the mutual inductances of the stator windings are considered small [10] and most researchers do not take them into account. For AFSRM this is not true because the magnetic flux paths are quite different to those found in radial flux SRM. In this work, the AFSRM self and mutual inductances were estimated based on the three-dimensional finite element method and then they were used in the motor electromechanical model.

When voltages or current pulses are imposed in the stator windings, the level of torque ripple is high, if compared with other kind of motors, which is the primary disadvantage of SRM as it contributes to the vibration and acoustic noise. For this reason, the imposition of different current shapes to reduce the torque ripple appears as an interesting solution. In this work, the procedure to obtain alternatives shapes is described and the effectiveness of this strategy is experimentally tested in an AFSRM prototype. Even more, in order to evaluate the performance of different kinds of speed controllers, three PID types were implemented: a conventional, a fuzzy logic and a neural based approach. Although in different levels the experimental results confirm the reduction of torque ripple, there are also differences with respect to the execution time and the speed response.

THE AXIAL FLUX SRM PROTOTYPE

The prototype has six poles in the stator, corresponding to three phases ("a", "b" and "c"), and four poles in the rotor. It is a 3-phase 6/4 poles AFSRM, shown in Figure 1.

The characteristics of the 3-phase 6/4 AFSRM prototype are shown in Table 1.

Table 1. The characteristics of the 3-phase 6/4 axial flux SRM

Parameter	Value
Outer diameter of rotor and stator	126 mm
Inner diameter of rotor and stator poles	63 mm
Shaft diameter	40 mm
Air gap	1.9 mm
Stator pole width	34 mm
Rotor pole width	26 mm
Stator and rotor poles arc	40°
Poles radial length	31.5 mm
Stator yoke thickness	5 mm
Stator pole area (axial cross-section)	1039 mm^2
Rotor yoke thickness	17 mm
Number of turns per stator pole	175
Turn wire	24 AWG
Coil resistance	2.3 Ω
Stator and rotor cores material	Steel SAE-1020
Motor shaft material	Stainless steel

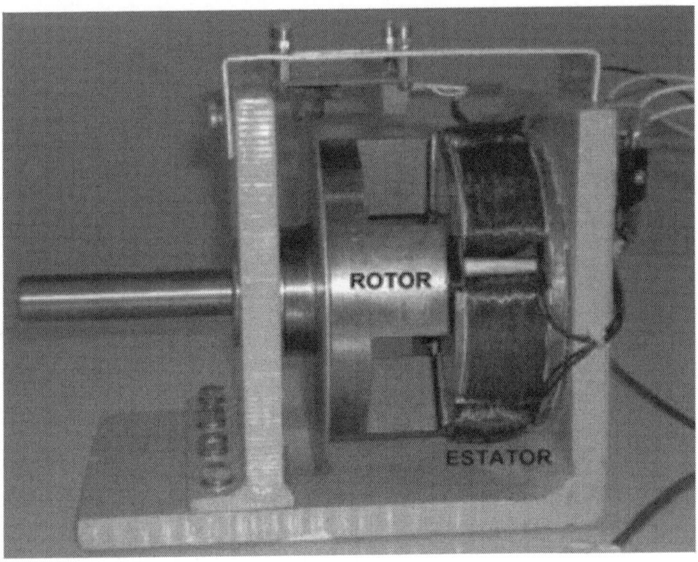

Figure 1. The AFSRM prototype.

Each stator pole has one coil, so there are six coils named "a1", "a2", "b1", "b2", "c1" and "c2".

The stator and rotor cores were solid because the objective of this work was the torque ripple minimization and not the loss reduction.

TORQUE RIPPLE MINIMIZATION STRATEGY

Considering that the motor core is operating in the linear region, the net electromagnetic torque T_e for the three phase AFSRM prototype is found as:

$$T_e(\theta_a, I_a, I_b, I_c) = \frac{1}{2}\begin{bmatrix} I_a(\theta_a) & I_b(\theta_a) & I_c(\theta_a) \end{bmatrix} \frac{\partial}{\partial\theta}\left(\begin{bmatrix} L_{11}(\theta_a) & L_{12}(\theta_a) & L_{13}(\theta_a) \\ L_{21}(\theta_a) & L_{22}(\theta_a) & L_{23}(\theta_a) \\ L_{31}(\theta_a) & L_{32}(\theta_a) & L_{33}(\theta_a) \end{bmatrix}\right) \begin{bmatrix} I_a(\theta_a) \\ I_b(\theta_a) \\ I_c(\theta_a) \end{bmatrix} \tag{1}$$

$$L_{11}(\theta_a) = +L_{a1a1}(\theta_a) + M_{a2a1}(\theta_a) + M_{a1a2}(\theta_a) + L_{a2a2}(\theta_a) \tag{2}$$

$$L_{12}(\theta_a) = +M_{b1a1}(\theta_a) - M_{b2a1}(\theta_a) - M_{b1a2}(\theta_a) + M_{b2a2}(\theta_a) \tag{3}$$

$$L_{13}(\theta_a) = -M_{c1a1}(\theta_a) + M_{c2a1}(\theta_a) + M_{c1a2}(\theta_a) - M_{c2a2}(\theta_a) \tag{4}$$

$$L_{21}(\theta_a) = +M_{a1b1}(\theta_a) - M_{a2b1}(\theta_a) - M_{a1b2}(\theta_a) + M_{a2b2}(\theta_a) \tag{5}$$

$$L_{22}(\theta_a) = +L_{b1b1}(\theta_b) + M_{b2b1}(\theta_b) + M_{b1b2}(\theta_b) + L_{b2b2}(\theta_b) \tag{6}$$

$$L_{23}(\theta_a) = +M_{c1b1}(\theta_b) - M_{c2b1}(\theta_b) - M_{c1b2}(\theta_b) + M_{c2b2}(\theta_b) \tag{7}$$

$$L_{31}(\theta_a) = -M_{a1c1}(\theta_a) + M_{a2c1}(\theta_a) + M_{a1c2}(\theta_a) - M_{a2c2}(\theta_a) \tag{8}$$

$$L_{32}(\theta_a) = +M_{b1c1}(\theta_b) - M_{b2c1}(\theta_b) - M_{b1c2}(\theta_b) + M_{b2c2}(\theta_b) \tag{9}$$

$$L_{33}(\theta_a) = +L_{c1c1}(\theta_c) + M_{c2c1}(\theta_c) + M_{c1c2}(\theta_c) + L_{c2c2}(\theta_c) \tag{10}$$

where: L and M stand for the self and mutual inductances.

The rotor position angle θ_a, for phase "*a*", is considered as 45° when the rotor pole was completely overlapping with the stator pole, as depicted in Figure 2, and this angle is measured clockwise.

The θ_b and θ_c angles correspond to rotor angular positions for phases "b" and "c", respectively. The relationships between these angles are:

- for $\theta_a \leq 30°$

$$\theta_b = \theta_a + 30^\circ \tag{11}$$

$$\theta_c = \theta_a + 60^\circ \tag{12}$$

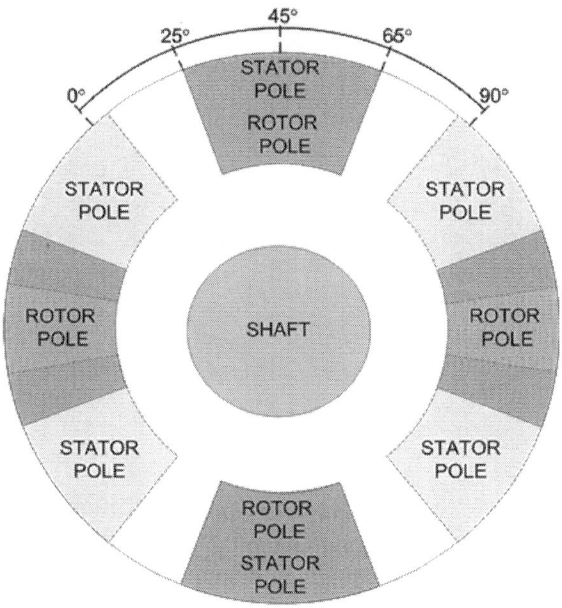

Figure 2. Position of rotor pole 45° relative to stator pole of the AFSRM prototype.

- for $30° < \theta_a \leq 60°$

$$\theta_b = \theta_a + 30^\circ \tag{13}$$

$$\theta_c = \theta_a - 30^\circ \tag{14}$$

- for $60° < \theta_a \leq 90°$

$$\theta_b = \theta_a - 60^o \tag{15}$$

$$\theta_c = \theta_a - 30^o \tag{16}$$

The relationship among the three phases self and mutual inductances are: $L_{a2a2}(\theta_a) = L_{a1a1}(\theta_a)$; $L_{b1b1}(\theta_b) = L_{a1a1}(\theta_b)$; $L_{b2b2}(\theta_b) = L_{b1b1}(\theta_b)$; $L_{c1c1}(\theta_c) = L_{a1a1}(\theta_c)$; $L_{c2c2}(\theta_c) = L_{c1c1}(\theta_c)$; $M_{a2a2}(\theta_a) = M_{a1a2}(\theta_a)$; $M_{a2b1}(\theta_a) = M_{a1b2}(\theta_a)$; $M_{a2b2}(\theta_a) = M_{a1b1}(\theta_a)$; $M_{a2c1}(\theta_a) = M_{a1c2}(\theta_a)$; $M_{a2c2}(\theta_a) = M_{a1c1}(\theta_a)$, $M_{b1a1}(\theta_a) = M_{a1b1}(\theta_a)$; $M_{b1a2}(\theta_a) = M_{a2b1}(\theta_a)$; $M_{b1b2}(\theta_b) = M_{a1a2}(\theta_b)$; $M_{b1c1}(\theta_b) = M_{a1b1}(\theta_b)$; $M_{b1c2}(\theta_b) = M_{a1b2}(\theta_b)$; $M_{b2a1}(\theta_a) = M_{a1b2}(\theta_a)$; $M_{b2a2}(\theta_a) = M_{a2b2}(\theta_a)$; $M_{b2b1}(\theta_b) = M_{b1b2}(\theta_b)$; $M_{b2c1}(\theta_b) = M_{b1c2}(\theta_b)$; $M_{b2c2}(\theta_b) = M_{b1c1}(\theta_b)$; $M_{c1a1}(\theta_a) = M_{a1c1}(\theta_a)$; $M_{c1a2}(\theta_a) = M_{a2c1}(\theta_a)$; $M_{b2b1}(\theta_b) = M_{b1b2}(\theta_b)$; $M_{c1b1}(\theta_b) = M_{b1c1}(\theta_b)$; $M_{c1b2}(\theta_b) = M_{b2c1}(\theta_b)$; $M_{c1c2}(\theta_c) = M_{a1a2}(\theta_c)$; $M_{c2a1}(\theta_a) = M_{a1c2}(\theta_a)$; $M_{c2a2}(\theta_a) = M_{a2c2}(\theta_a)$; $M_{c2b1}(\theta_b) = M_{b1c2}(\theta_b)$; $M_{c2b2}(\theta_b) = M_{b2c2}(\theta_b)$; and $M_{c2c1}(\theta_c) = M_{c1c2}(\theta_c)$.

The self and mutual inductances were estimated based on a simulation model using a three dimensional finite element method of the "ANSYS Multiphysics" software [11] with magneto static analysis. The 96 type solid elements were used in the simulation software.

In Figure 3, for eleven rotor positions, the self and mutual inductances estimated, for phase "a", are shown. The self-inductance is represented by L_{a1a1} and the mutual inductances are named M_{a1a2}, M_{a1b1}, M_{a1b2}, M_{a1c1} and M_{a1c2}, respectively.

As shown in Figure 3, the self-inductance has a similar profile to the radial flux SRM case, which means that the minimum inductance corresponds to the unaligned position (no overlap between rotor and stator poles) while the maximum inductance corresponds to the full alignment position between rotor and stator poles. The inductances profiles repeat every 90°.

From Figure 3, it can be noted that the mutual inductances between the coil "a1" and the other five coils are not negligible. For instance, the maximum relationship between them (M_{a1b1}/L_{a1a1}) is around 0.25 near 20°.

In this way, in order to provide a constant electromagnetic torque for all rotor positions, appropriated reference currents should be designed. From previous tests in open loop [12] the best current feed strategy consists of the following sequence of energizing the windings:

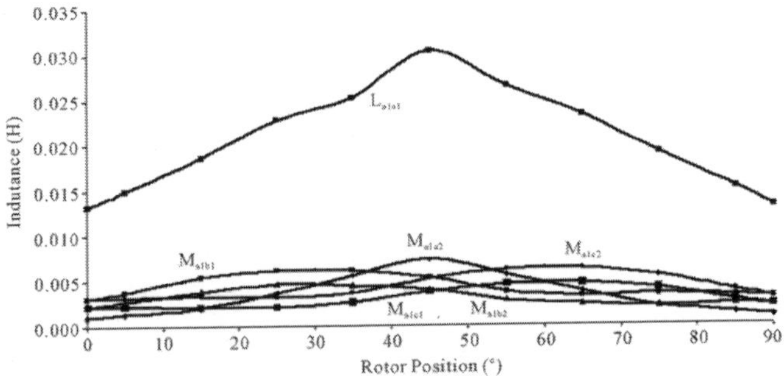

Figure 3. AFSRM self and mutual inductances.

- phases "a" and "b" from rotor position 0° to 15°;
- only phase "a" from rotor position 15° to 30°;
- phases "a" and "c" from rotor position 30° to 45°;
- only phase "c" from rotor position 45° to 60°;
- phases "b" and "c" from rotor position 60° to 75°;
- only phase "b" from rotor position 75° to 90°.

This cycle repeats each 90°.

On the other hand, to define a reference torque without harmonics, it was considered as that generated when a current of 3A is imposed in phase "a", while the rotor position was 30°. In this situation a net electromagnetic torque of 0.126 Nm was calculated.

Next, for every one degree, the current values of the phase "a" were calculated as:

- from rotor position 15° to 30°

$$I_a(\theta_a) = \sqrt{\frac{0.126}{\frac{\partial L_{a1a1}(\theta_a)}{\partial \theta_a} + \frac{\partial L_{a1a2}(\theta_a)}{\partial \theta_a}}} \quad (17)$$

- from rotor position 0° to 15° a function that starts in zero and it then assumes the value calculated with (17) in the rotor position of 15°, was used

$$I_a(\theta_a) = 2.7967 - 2.7967e^{(-\theta_a/4)} \quad (18)$$

- from rotor position 30° to 45° the I_a current values are calculated from quadratic Equation (19) considering that I_c assumes the values of I_a currents obtained from (18). For example, the I_c value for 35° is equal to the I_a value for 5°

$$\left(\frac{\partial L_{a1a1}(\theta_a)}{\partial \theta_a}+\frac{\partial L_{a1a2}(\theta_a)}{\partial \theta_a}\right)I_a^2(\theta_a)+\left[\left(\frac{-2\partial L_{a1c1}(\theta_a)}{\partial \theta_a}+\frac{2\partial L_{a1c2}(\theta_a)}{\partial \theta_a}\right)I_c(\theta_a)\right]I_a(\theta_a)$$
$$+\left(\frac{\partial L_{c1c1}(\theta_a)}{\partial \theta_a}+\frac{\partial L_{c1c2}(\theta_a)}{\partial \theta_a}\right)I_c^2(\theta_a)-0.126=0 \tag{19}$$

The reference currents of phases "c" and "b" are shifted by 30° and 60° relative to phase current "a". All these reference currents are shown in Figure 4.

CONTROLLERS DESCRIPTION

In order to evaluate the performance of different kinds of speed controllers, three PID types were implemented: a conventional, a fuzzy logic and a neural based approach.

In Figure 5, the conventional PID structure is shown. As noted, the current references are obtained multiplying the controller output by a table with three outputs, which depend on the angular position of the rotor. In our experiments, two tables were evaluated. In the first one, the outputs have pulsed shapes. The final effect is a torque with high ripple components. In the second case, the outputs have modified shapes according to Figure 4. As the experimental results show, the reduction of torque ripple is important.

In this case, the PID gains were adjusted by simulations using MatLab® Simulink software, resulting in gains proportional, integral and derivative, of 16, 3 and 1, respectively.

The controller output was limited to the interval of [0, 1].

In Figure 6, the fuzzy or neural control structure is shown. As noted, the outputs of the controller are directly the current references for the AFSRM.

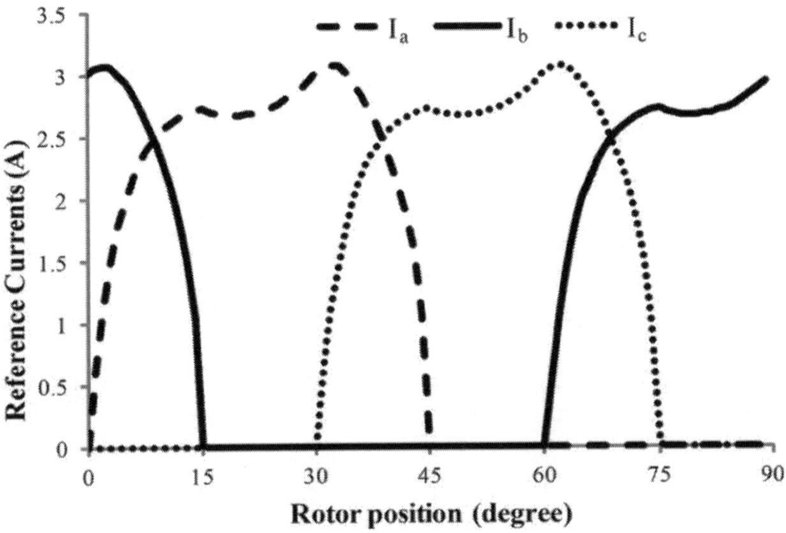

Figure 4. Reference currents waveforms.

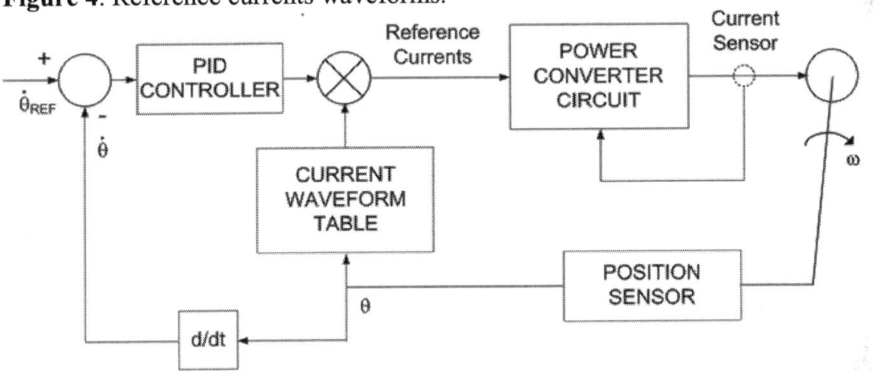

Figure 5. Block diagram of the PID speed control system.

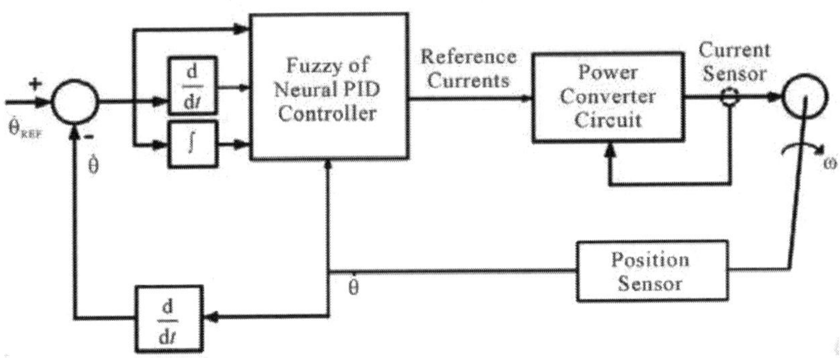

Figure 6. Block diagram of the fuzzy or neural PID speed control system.

In the case of the fuzzy controller, the rules were based on the observation of the simulation results of the PID controller that use the modified current shapes. The inputs of the fuzzy logic controller were speed error, derivative of speed error, integral of speed error and angular rotor position. The output variables are the three phase's reference currents. In the actual implementation, this strategy was translated to a table.

The fuzzy logic characteristics used in MatLab® Simulink simulations are: minimum "And" Method, maximum "Or" Method, minimum "Implication", maximum "Aggregation" and centroid "Defuzzification".

The input variable speed error ("E") has three linguistic values: "negative", "zero" and "positive". The corresponding membership functions are:

- trapezoidal type $[-50 \ \ -50 \ \ -1 \ \ 0]$ for "negative" ("N");
- triangular type $[-1 \ \ 0 \ \ 1]$ "zero" ("Z");
- trapezoidal type $[0 \ \ 1 \ \ 350 \ \ 350]$ for "positive" ("P").

The input variable derivative of speed error ("CE") has three linguistic values: "negative", "zero" and "positive". The corresponding membership functions are:

- triangular type $[-20 \ \ -20 \ \ 0]$ for "negative" ("N");
- triangular type $[-20 \ \ 0 \ \ 20]$ for "zero" ("Z");
- trapezoidal type $[0 \ \ 20 \ \ 20]$ for "positive" ("P").

The input variable integral of speed error ("IE") has three linguistic values: "negative", "zero" and "positive". The corresponding membership functions are:

- triangular type $[-10 \ \ -10 \ \ 0]$ for "negative" ("N");
- triangular type $[-10 \ \ 0 \ \ 10]$ for "zero" ("Z");
- triangular type $[0 \ \ 10 \ \ 10]$ for "positive" ("P").

The input variable angular rotor position ("Pos") has ten linguistic values: "P1", "P2", "P3", "P4", "P5", "P6", "P7", "P8", "P9" and "P10", whose corresponding membership functions are shown inFigure 7.

The output variables are the three phase's reference currents. They have four linguistic values: "zero" ("Z"), "low" ("L"), "medium" ("M") and "high" ("H"). The corresponding membership functions for phase "a" current are shown in Figure 8.

Take into account the number of membership functions, there are 270 (3 × 3 × 3 × 10) fuzzy rules. Its structure is similar to: if (E is "P") and (CE is "P") and (IE is "N") and (Pos is "P5") then (I_a is "M") (I_b is "Z") (I_c is "H").

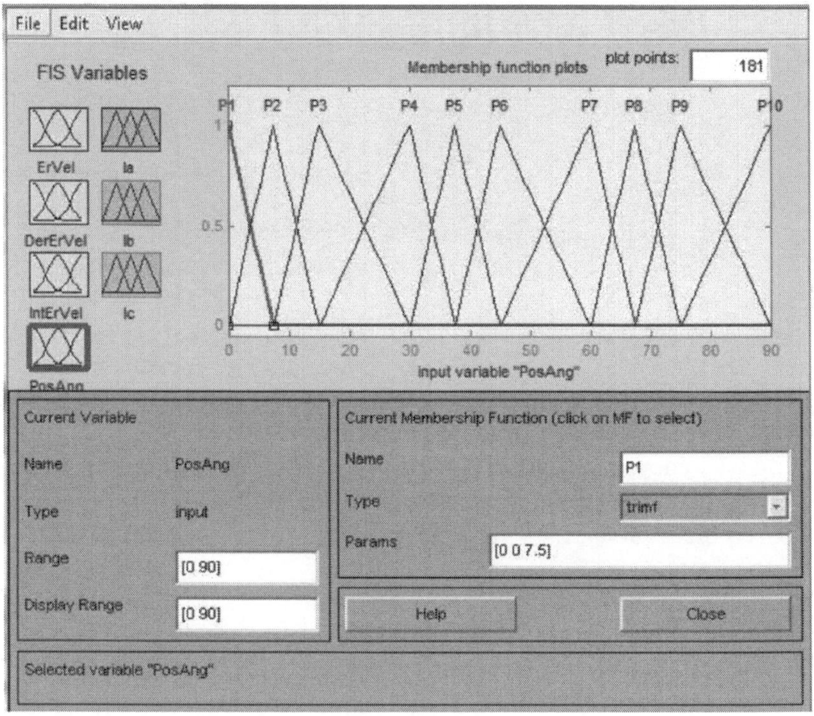

Figure 7. Membership functions of input variables of the fuzzy speed controller.

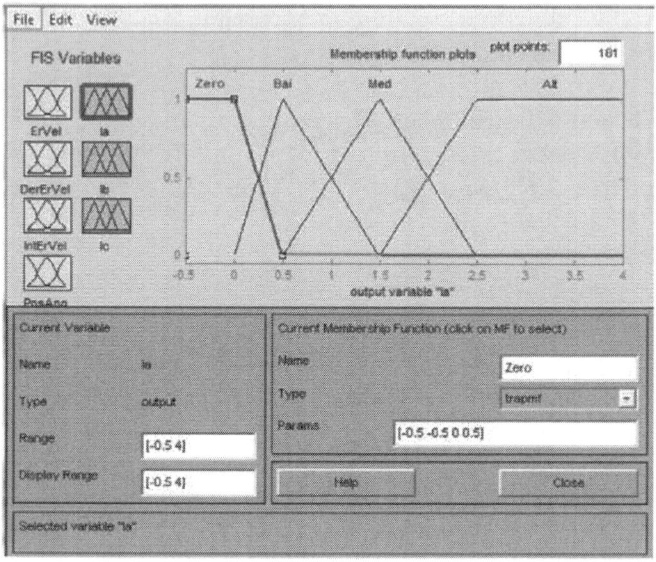

Figure 8. Membership functions of output variables of the fuzzy speed controller.

For the neural PID controller design, the usual procedure of off-line training was adopted. This was based in the data, with sampling frequency of 1 kHz, from the simulation results of the PID controller that use the modified current shapes.

The simulation with MatLab® Simulink used 12,000 input data values (speed error, derivative of speed error, integral of speed error and angular rotor position) distributed as 6000 for training, 3000 for test and 3000 for validation. On the other hand, 12,000 output data values (currents in three phases) were used, obtained with the simulation of the PID conventional controller.

For neural network training the MatLab® software was used ("nntool"). The neural network properties were: feed-forward back propagation network type, network training function that updates weight and bias values according to Levenberg-Marquardt optimization (TRAINLM), gradient descent with momentum weight and bias adaption learning function (LEARNGDM), mean squared error performance function (MSE), two layers: layer 1 using 36 neurons with hyperbolic tangent sigmoid transfer function (TANSIG) and layer 2 using 3 neurons with linear transfer function (PURELIN). Figure 9 shows the neural architecture and Figure 10 shows the performance of the neural network training used in the simulation.

The training generated weight and bias values of the two layers that were used in the neural PID controller to calculate the references currents of the simulation. In the actual implementation, a correspondent table was prepared.

CONTROLLERS IMPLEMENTATION

The hardware to control the AFSRM consists of a controller circuit, three power converter circuits, one for each phase, a position sensor circuit, three Hall Effect current measurement circuits, also one for each phase, a torque meter connected between the AFSRM shaft and the mechanical load shaft. The shaft of a DC machine was used as an inertial load. These components are arranged as shown in the block diagram of Figure 11.

The experimental hardware is shown in Figure 12, including the AFSRM. In this photo the parts of the hardware used in the experiments can be identified: the controller circuit (1), the power converter circuits (2), the position sensor circuit (3), Hall Effect current measurement circuits (4), torque sensor and display (5), the AFSRM (6) and the DC machine (7).
The position sensor consists of an infrared optical circuit and an aluminum dish with 180 holes, fixed to the rotor structure. The signal from the sensor is transformed into a rectangular pulse stream with amplitude of 5 V, which is sent to the controller circuit.

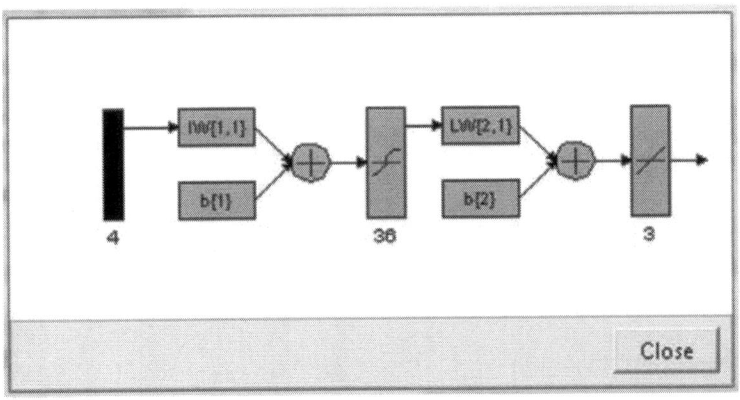

Figure 9. Neural architecture generated by “nntool” of the MatLab®software.

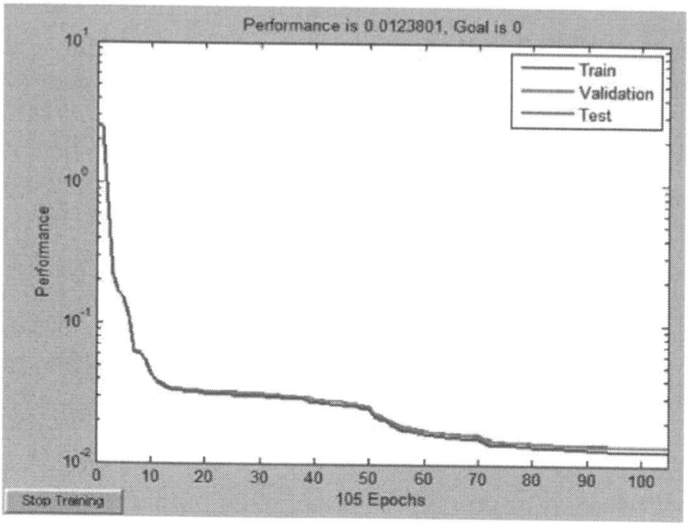

Figure 10. Neural network training of the MatLab® software.

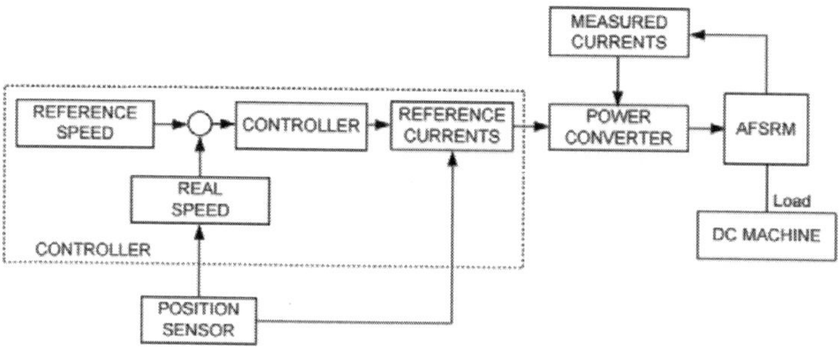

Figure 11. AFSRM drive block diagram.

The actual phase currents are measured using commercial Hall Effect circuits. Their gain and zero adjustments were done in order to obtain 1 V for 1 A of phase current.

The power circuits consist of three asymmetric bridge converters and auxiliary circuits. In Figure 13, one of the power circuits is depicted.

As illustrated, the actual measured current is compared with the reference current coming from the controller circuit. If the real current is greater than the reference, the power circuit switches off the Mosfets of this phase in order to decrease the current in the coils.

The controller circuit consists of one microcontroller PIC 18F4680, programmed using the "C" language, that receives the rectangular pulses from the position sensor and calculates the reference currents of the three phases in accordance with the controller type used, sending them to the power converter. Due to the 180 holes in the aluminum dish, the controller imposed the reference currents every 2°.

Figure 12. Experimental hardware and AFSRM prototype.

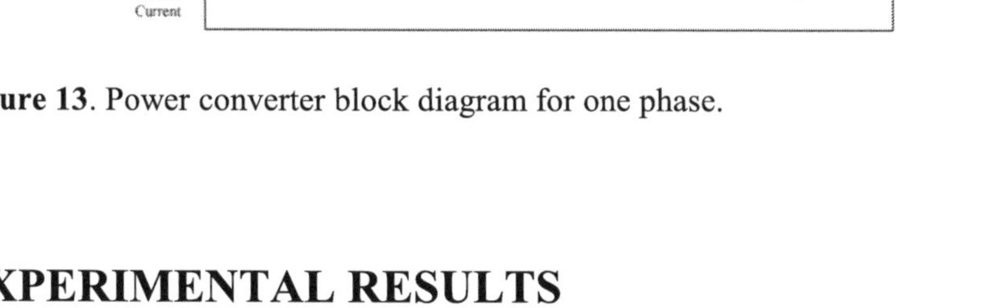

Figure 13. Power converter block diagram for one phase.

EXPERIMENTAL RESULTS

In order to evaluate the torque ripple minimization strategy, four experiments were performed:

- current pulses were imposed and a conventional PID speed controller was used;

- proposed current waveforms, shown in Figure 4, were imposed and a conventional PID speed controller was used;
- a fuzzy logic PID speed controller was used;
- a neural PID speed controller was used.

Imposing Current Pulses

In this case the current pulses were applied in the following sequence:

- phase "a" from rotor position 0° to 30°;
- phase "c" from rotor position 30° to 60°;
- phase "b" from rotor position 60° to 90°.

The Figure 14 shows the reference current imposed by the controller and the actual current in the coils of one phase.

The Figure 15 shows the speed response and the Figure 16 shows the torque response, when current pulses are imposed and a conventional PID speed controller is used. The reference speed is 350 rpm.

In this case, the torque ripple in steady state was around 17.1% of the average torque.

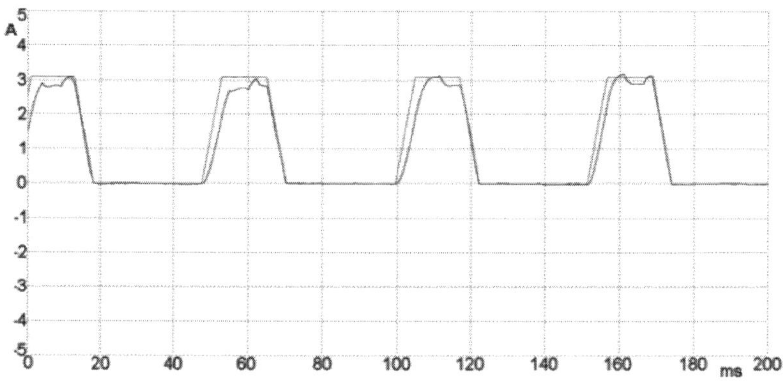

Figure 14. Reference pulsed current and real current.

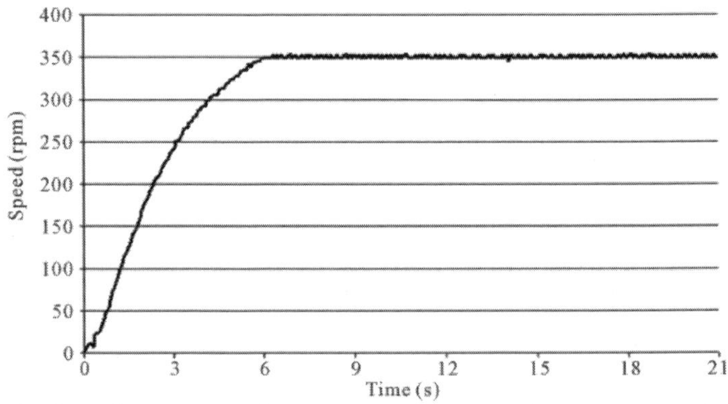

Figure 15. Speed response for current pulses and PID conventional speed controller.

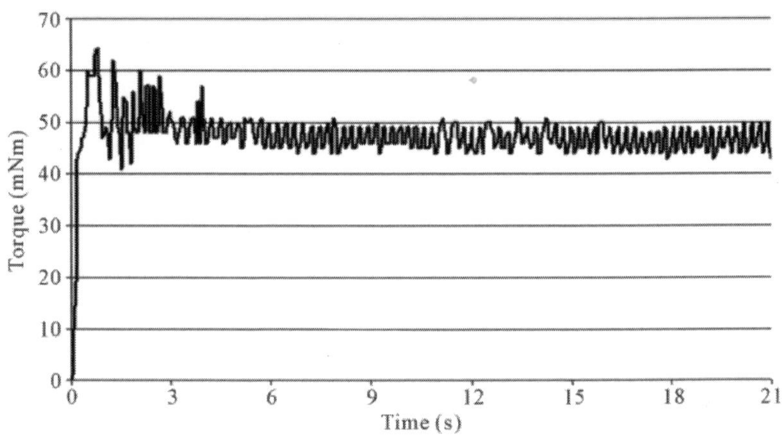

Figure 16. Torque response for current pulses and conventional PID speed controller.

Imposing Proposed Current Waveforms

The current waveform proposed as reference and the actual current in phase "a" are shown in Figure 17.

Note that the actual and reference currents are practically overlapping.
The Figure 18 shows the speed response and Figure 19 shows the torque response for current waveforms proposed and conventional PID speed controller.

In this case, the torque ripple in steady state was around 2.1% of the average torque.

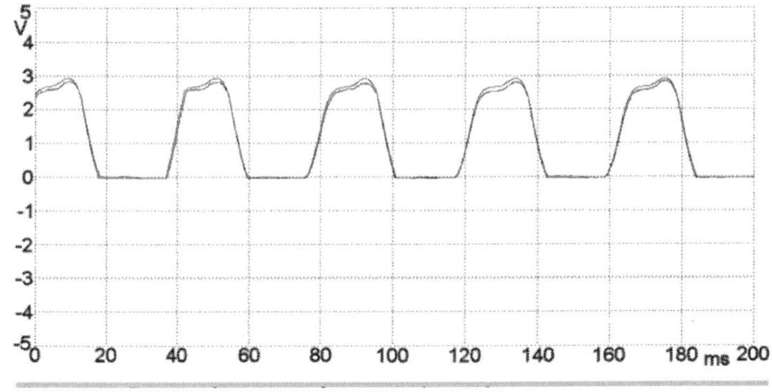

Figure 17. Proposed current reference and actual current.

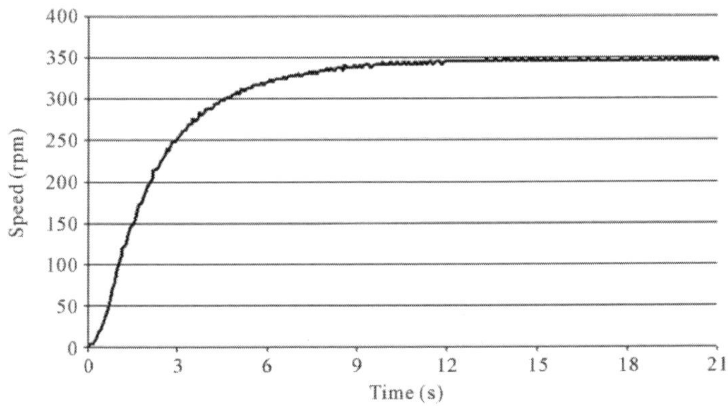

Figure 18. Speed response for current waveforms proposed and conventional PID speed controller.

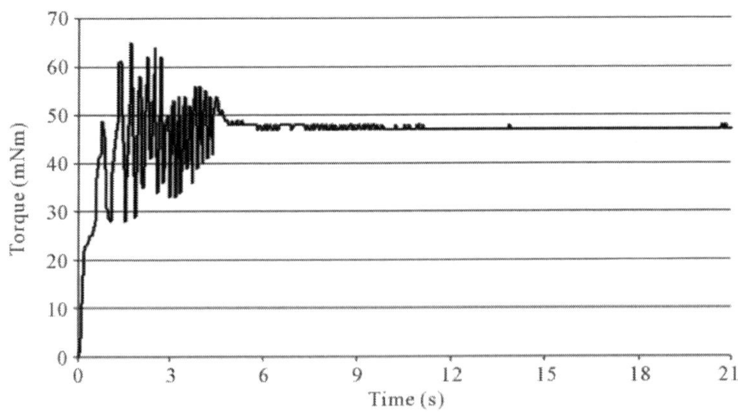

Figure 19. Torque response for current waveforms proposed and conventional PID speed controller.

Using Fuzzy Logic PID Speed Controller

The actual currents in two phases of the AFSRM are shown in Figure 20. These were very close to their references.

Figure 21 shows the speed response and Figure 22 shows the torque response for the fuzzy logic PID controller.

In this case, the torque ripple in steady state was around 6.5% of the average torque.

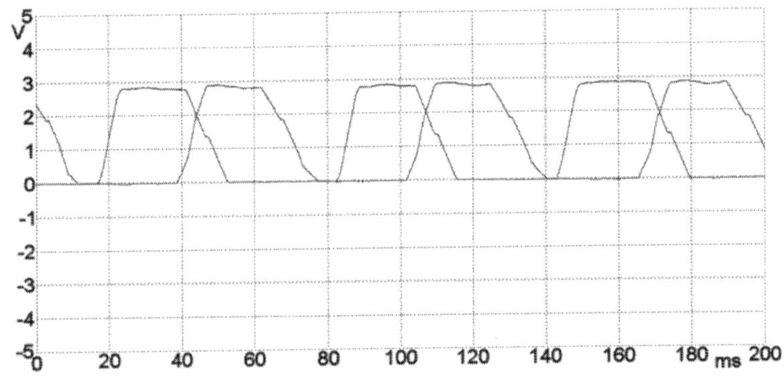

Figure 20. Actual currents of two phases using the fuzzy logic PID controller.

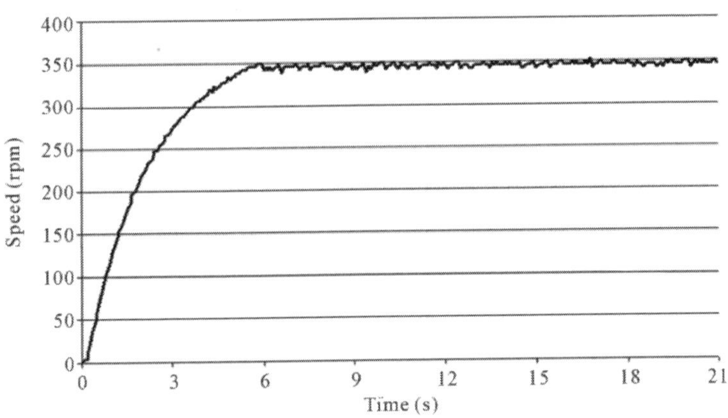

Figure 21. Speed response for the fuzzy logic PID controller.

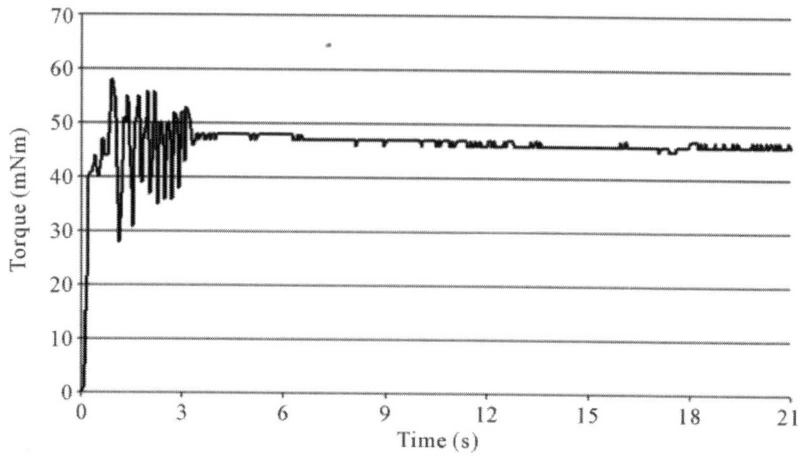

Figure 22. Torque response for the fuzzy logic PID controller.

Using Neural PID Speed Controller

The actual currents in two phases of the AFSRM are shown in Figure 23.

The Figure 24 shows the speed response and the Figure 25 shows the torque response for neural controller.

In this case, the torque ripple in steady state was around 2.0% of the average torque.

CONCLUSIONS

In this work, particular reference waveforms for the currents of an ASFRM in a closed loop speed control

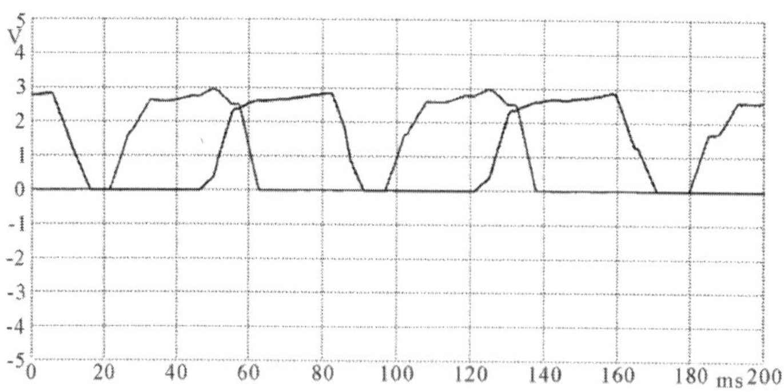

Figure 23. Actual currents of two phases using the neural PID controller.

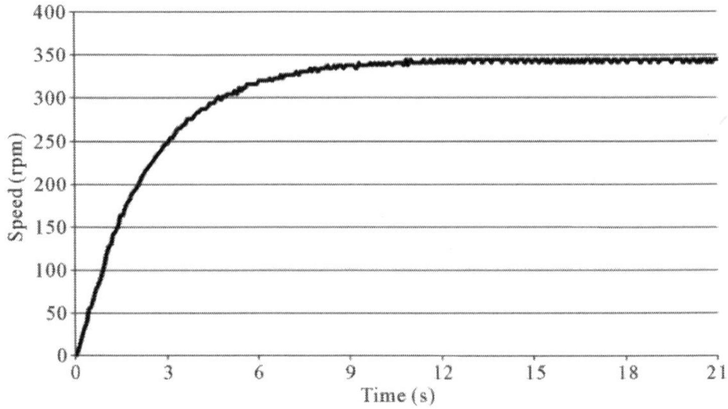

Figure 24. Speed response for neural PID controller.

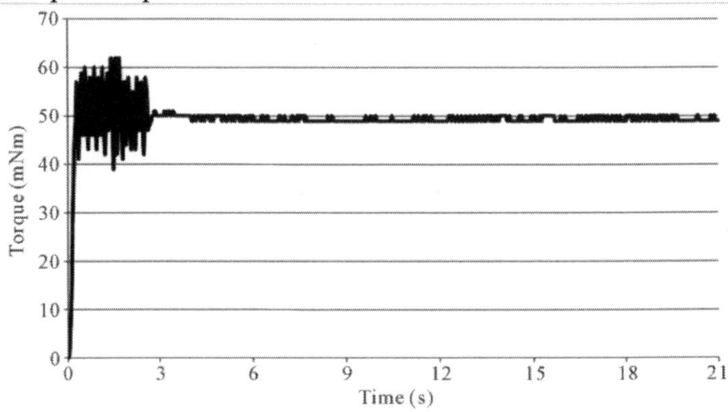

Figure 25. Torque response for neural PID controller.

system have been proposed and experimentally tested. Comparing the performance of a conventional PID speed controller using current pulses with a conventional PID speed controller, using the current waveforms proposed, it is confirmed that the level of torque ripple is reduced.

All the controllers were adjusted through simulations, using a motor model that is composed by self and mutual inductances obtained through 3D FEM simulations.

Comparing the performance of a conventional PID speed controller using the proposed current waveforms with a fuzzy logic PID speed controller, it was noted that the torque ripple for the first one is lower, which can be explained because the currents obtained with fuzzy logic controller are not exactly similar to those that would produce constant torque.

Nevertheless, in respect of the speed response, the fuzzy logic PID controller reaches the steady state in less time than the conventional PID controller. Even more, with respect to the implementation, the execution time of the fuzzy logic PID is lower than that used with the conventional PID controller but similar to that spent by the neural PID controller.

Finally, with respect to the neural PID controller, it notes similar torque ripple to the conventional PID controller but lower than that obtained using the fuzzy logic approach, which is expected as this was obtained from training based on the conventional PID controller results. However, this has a cost, which is the time spent in off-line training.

REFERENCES

1. Krishnan, R. (2001) Switched Reluctance Motor Drives: Modeling, Simulation, Analysis, Design and Applications. CRC Press, Boca Raton.http://dx.doi.org/10.1201/9781420041644
2. Miller, T.J.E. (1993) Switched Reluctance Motors and Their Control. Oxford Science,Oxford .
3. Vijayakumar, K., Karthikeyan, R., Paramasivam, S., Arumugam, R. and Srinivas, K.N. (2008) Switched Reluctance Motor Modeling, Design, Simulation, and Analysis: A Comprehensive Review. IEEE Transactions on Magnetics, 44, 4605-4617.http://dx.doi.org/10.1109/TMAG.2008.2003334
4. Huang, S., Luo, J., Leonardi, F. and Lipo, T.A. (2000) A Comparison of Power Density for Axial Flux Machines Based on General Purpose Sizing Equations. IEEE Transactions on Energy Conversion, 14, 185-192. http://dx.doi.org/10.1109/60.766982
5. Madhavan, R. and Fernandes, B.G. (2013) Axial Flux Segmented SRM with a Higher Number of Rotor Segments for Electric Vehicles. IEEE Transactions on Energy Conversion, 28, 203-213. http://dx.doi.org/10.1109/TEC.2012.2235068
6. Madhavan, R. and Fernandes, B.G. (2012) Comparative Analysis of Axial Flux SRM Topologies for Electric Vehicle Application. IEEE International Conference on Power Electronics, Drives and Energy Systems, Bengaluru, 16-19 December 2012, 1-6.
7. Asghar, K. (2013) Analysis of Switched Reluctance Motor Drives for Reduced Torque Ripple Using FPGA Based Simulation Technique. American Journal of Information Sciences, 6, 1-11.
8. Wadnerkar, V.S., Bhaskar, M.M., Das, T.R. and RajKumar, A.D. (2010) A New Fuzzy Logic Based Modeling and Simulation of a Switched Reluctance Motor. Journal of Electrical Engineering & Technology, 5, 276-281. http://dx.doi.org/10.5370/JEET.2010.5.2.276

9. Raj, E.F.I. and Kamaraj, V. (2013) Neural Network Based Control for Switched Reluctance Motor Drive. IEEE International Conference on Emerging Trends in Computing, Communication and Nanotechnology, 678-682.
10. Paula, P.P., Silva, W.M., Cardoso, J.R. and Nabeta, S.I. (2003) Assessment of the Influences of the Mutual Inductances on Switched Reluctance Machines Performance. International Electric Machine and Drives Conference, 3, 1697-1701.
11. ANSYS Inc. (2005) Low-Frequency Electromagnetics Analysis Guide: Release 10.0. ANSYS Inc., Canonsburg.
12. Sass, F., Santisteban, J.A. and Sanches, E. (2009) Design and Implementation of a Digital Control System for an Axial Flux Switched Reluctance Motor. Brazilian Power Electronics Conference, Bonito, 27 September-1 October 2009, 138-144.

CITATION

Eric S. Sanches, José A. Santisteban, Comparative Study of Conventional, Fuzzy Logic and Neural PID Speed Controllers with Torque Ripple Minimization for an Axial Magnetic Flux Switched Reluctance Motor, DOI: 10.4236/eng.2014.611065

CHAPTER 9

Design of Real Time Battery Management Unit for PV-Hybrid System by Application of Coulomb Counting Method

Apiwat Ausswamaykin[1], *Boonyang Plangklang*[2]

[1]Department of Electrical Engineering, Faculty of Engineering
Rajamangala University of Technology IsanKhonkaen Campus, Khon Kaen, Thailand
[2]Department of Electrical Engineering, Faculty of Engineering
Rajamangala University of Technology Thanyaburi, Pathum Thani, Thailand

ABSTRACT

This paper presents a real-time battery management unit designed by applying the Coulomb counting method and intended for use in an integrated renewable energy system for PV-Hybrid power supply. Battery management is required to stabilize hybrid systems and extend battery lifetimes. The battery management unit is divided into three main stages. Firstly, analysis of the basic components of the battery type used in the system is considered. Secondly, the state of charge (SOC) estimation method and the deterioration factor of the battery are analyzed. Finally, the overall battery management system, including a computer-based measurement and control unit, is constructed. The control system displays real-time information through LabVIEW 8.5 by estimating the state of charge through various measurements. The system will issue alerts when malfunctions are detected, and the operator can analyze and react to the system in real time to stabilize the system and extend the battery lifetime.

INTRODUCTION

Electrical energy demand increases every year and most electrical energy production begins with coal, natural gas and petroleum, which affect the environment through pollution. Therefore, electrical energy from a hybrid renewable energy source has become an attractive alternative for supplying electrical energy.

Hybrid renewable energy systems (HRES) are becoming popular for power generation in remote areas due to advances in renewable technologies and increasing petroleum products prices. A hybrid energy system generally consists of two or more renewable energy sources used together to increase the system's efficiency and balance the energy supply. The most frequently used type of HRES is a photovoltaic (PV) array coupled to a generator. Hybrid energy systems often yield greater economic and environmental returns than stand-alone photovoltaic systems. To stabilize the HRES, an energy storage system is required to supplement the generator when the PV cannot generate electricity, e.g., at night or under cloud coverage. The normal HRES energy storage unit is a battery system. The primary concerns with these battery storage systems are performance, lifetime and durability. Therefore, in these applications, a battery management system (BMS) is needed to control the working conditions of the battery, stabilizing the system and extending the useful life of the battery.

BASIC ELEMENTS OF A BATTERY IN SOLAR ENERGY SYSTEM

In practice, all types of battery can be used in renewable energy systems. We consider lead-acid, nickel-cadmium, and nickel-iron batteries and find that the battery type most commonly used in renewable energy systems is the deep-discharge lead-acid battery. Lead-acid batteries are preferred because of factors including the price, performance and battery lifetime, as shown in Table1

THEORY OF THE LEAD-ACID BATTERY [2]

The lead-acid battery is frequently used in automobiles and other applications in which weight is not a concern. In the 1970s, the valve-regulated lead-acid battery (or the "sealed lead-acid battery") was developed; this battery used a gel electrolyte instead of a liquid, allowing the battery to be used in a variety of positions without introducing leakage. In the battery's discharged state, both the positive and negative plates are lead sulfate ($PbSO_4$), and the electrolyte is primarily water, having lost most of its dissolved sulfuric acid (H_2SO_4). The discharge process is driven by the conduction of electrons from the negative plate to the positive plate through an external circuit. In the charged state, each cell contains negative plates of elemental lead (Pb) and positive plates of lead oxide (PbO_2) in a sulfuric acid electrolyte.

The charging process is driven by the forcible removal of electrons from the positive plate and the forcible introduction of them to the negative plate using a charging source. There are several types of lead-acid batteries, and the type most suited for use with renewable energy systems is the deep discharge type because this design is intended to provide small amounts of power continuously over long periods and discharge up to 80% of the total battery capacity without damaging the battery[3] [4] .

STATE OF CHARGE ESTIMATION METHOD

The state of charge (SOC) is the equivalent of a fuel gauge for a battery pack. The units of SOC are percentage points: 0 percent is empty, and 100 percent is full. An alternate form of the same measure is the depth of discharge (DOD), which is one minus the SOC (i.e., 100 percent is empty, 0 percent is full). The SOC is commonly used to discuss the current state of a battery in use, while the DOD is most often used when discussing the lifetime of the battery after repeated use.

State estimation must be accurate in a battery management system, as the system controls the performance of the battery to extend its lifetime. In practical applications, estimating the SOC is problematic because the load on a hybrid system is not smooth or stable.

Table 1. Comparisons of three battery types [1]

Type	Cycle life until 80% DOD	Investment cost [€/kWh]	Specific kWh cost [€/kWh$_S$]	h_T [%]	Self-discharge [%/month]	Temp. range [°C]
Pb	500...1500	85...350	0.17...0.30	>80	3...4	−15°...+50°
NiCd	1500...3500	650...1500	0.30...1.00	71	6...20	−40°...+45°
NiFe	3000	1000	0.33	55	40	0°...+40°

It is therefore necessary to estimate the state of charge of a battery for use in a battery management system (BMS) designed to increase the performance and reliability of the battery.

Many methods of estimating the state of charge exist. Table 2 presents a comparison of SOC estimation techniques.

The paper presents a Coulomb counting SOC estimation method for application in PV-hybrid power supply battery management systems because this method can be monitored in real-time under unstable load consumption conditions and is easy to calculate.

SOC ESTIMATION BY COULOMB COUNTING

The state of charge is estimated by measuring the battery current and integrating it over time. However, this method suffers from long-term drift and lack of a reference point. Therefore, the SOC must be re-calibrated on a regular basis, e.g., by resetting the SOC to 100 percent when a charger determines that the battery is fully charged.

For real-time estimation, this method utilizes Equation (1) with real-time measurements from the sensors and an initial SOC of 100 percent [6]-[11].

$$SOC(t) = SOC_{init} - \int_0^t \frac{i(t)}{C_n} dt \quad (1)$$

Where

$SOC(t)$ is Real-time state of charge (%).

SOC_{init} is Initial state of charge (%).

$i(t)$ is Real-time discharge current (A).

C_n is Capacity of battery (Ah).

DETERIORATION FACTOR OF BATTERY [12]

Ambient temperature effects: The lead-acid battery consists of a positive lead oxide plate and a negative lead plate. Both plates are dipped in sulfuric acid, and the temperature affects the rates of corrosion and deterioration of the plates. Temperature specifications are given by IEEE1184-1994[13].

Float charge voltage: A charging voltage different from the target voltage will affect the deterioration of the plate. When the floating charge voltage is under the target, the battery will experience sulfating at the positive plate and negative plate. If the floating charge voltage is over the target, then the battery will lose water.

Discharge cycle: Repeated discharge cycles cause battery deterioration. The battery plates will exhibit corrosion due to battery deterioration.

PROPOSED BATTERY MANAGEMENT SYSTEM [14] [15]

This section presents the basic theory of the lead-acid battery, techniques for estimating the state of charge and means of determining the deterioration state of the battery. The battery management system for PV-hybrid power supplies is summarized in Figure 1 and Figure 2.

Table 2. Example SOC estimation techniques [5]

Technique	Battery type	Advantages	Drawbacks
Discharge test	All battery system	Easy and accurate	Offline, time intensive, loss of energy
Voltage Based Estimation	Lead, Lithium, Zn	Easy	Long rest time, offline
Coulomb Counting	All battery system	Online, easy	Needs re-calibration point
Kalman Filtering	All battery system	Online, Dynamic	Computationally intensive, Needs a suitable battery model
Neural Network	All battery system	Online	Need training data of similar battery

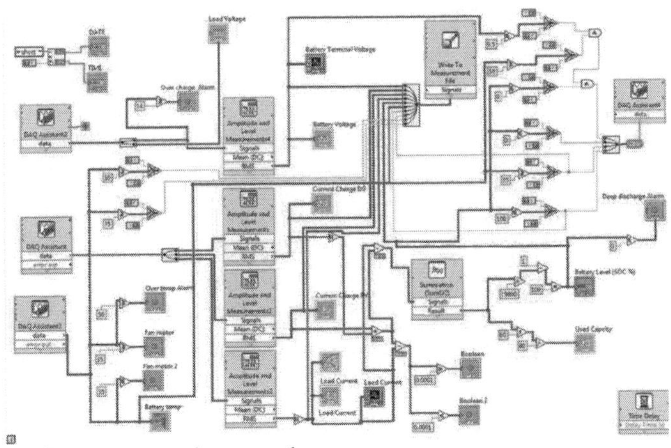

Figure 1. BMS programmatic control.

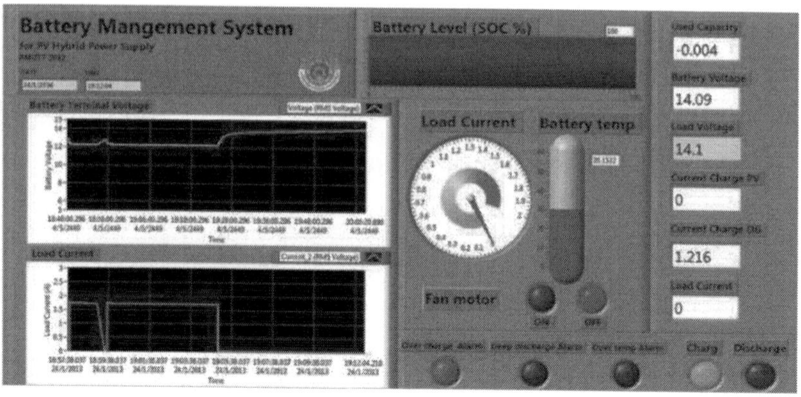

Figure 2. BMS monitoring display.

Figure 1 shows the program for battery control and SOC calculation. Figure 2 presents the monitoring display of the battery management system, which comprises 4 sections. Section 1 presents graphs of the voltage and load current of the battery. Section 2 shows digital values indicating the battery voltage, PV voltage, PV charge current and DC generator charge current. Section 3 presents a graphic display of the battery SOC. Finally, Section 4 issues alerts if the system detects malfunctions so that the operator can analyze and react to the system in real-time.

Data Acquisition

All algorithms that the BMS uses take measured data as input information. Therefore, the accuracy, sampling rate and characteristics of front-end filtering are crucial to the BMS and application-dependent.

The BMS also handles communication between the BMS and other onboard and off board devices. Depending on the application, different interface system may be used for data exchange. This paper utilized a Local Area Network: LAN by Nic DAQ-9188 and module NI9335, NI9227, NI9211 and NI9401, as shown in Figure 3.

Electrical Management

The input parameters of the electrical management system are the current, voltage and SOC. This system controls the charging and discharging processes.

Thermal Management

A thermal management is necessary for most high-power applications and high-temperature batteries. The thermal management system equalizes the temperature of the battery and cools it when necessary. Liquid or airbased thermal systems are used. This paper presents an air-base battery cooling system, with the battery operated at 35˚C and a stop condition at 50˚C.

Safety Management

The safety management system protects the battery against critical operation conditions. The design constraints employed in this paper are:

- The deep discharge cycle control criterion stops battery operation when the battery voltage reaches 80% DOD.
- The overcharge voltage control criterion controls the battery voltage to not exceed 115 percent of the rate voltage.

Battery State Determinations

The battery state is a crucial input parameter for the management system. For SOC determination in this paper employs the Coulomb counting method to determine the SOC of the battery. The results of the BMS employing the Coulomb counting method will be compared with standard tests performed on MTB-series module (Machine for Test Battery, MTB) at the RMUTT Laboratory as in Figure 4 and Figure 5 using lead-acid batteries supplying 12 V and 5.5 Ah. The discharge current is 1A. The results of this comparison are shown in Figure 6.

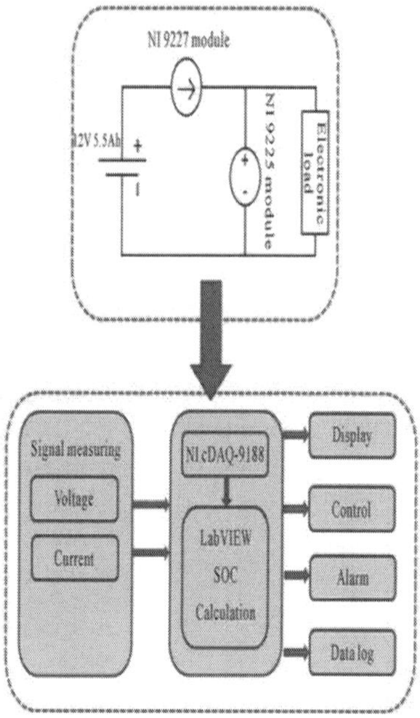

Figure 3. Data acquisition block of the Coulomb counting method.

employs the Coulomb counting method to determine the SOC of the battery. The results of the BMS employing the Coulomb counting method will be compared with standard tests performed on MTB-series module (Machine for Test Battery, MTB) at the RMUTT Laboratory as in Figure 4 andFigure 5 using lead-acid batteries supplying 12 V and 5.5 Ah. The discharge current is 1A. The results of this comparison are shown in Figure 6.

Figure 6 compares the SOC of the two systems, starting from 100 percent. The Coulomb counting method has 10 percent error relative to the MTB-series estimation.

CONCLUSIONS

This paper presents a real-time battery management system for use with PV-hybrid power supplies. The BMS employs the Coulomb counting method.

The designed system has a protection system, a control system and a monitoring system. Data acquisition from the battery is performed using LabVIEW8.5 and the Nic DAQ-9188 module and serves as the input parameters to the other systems.

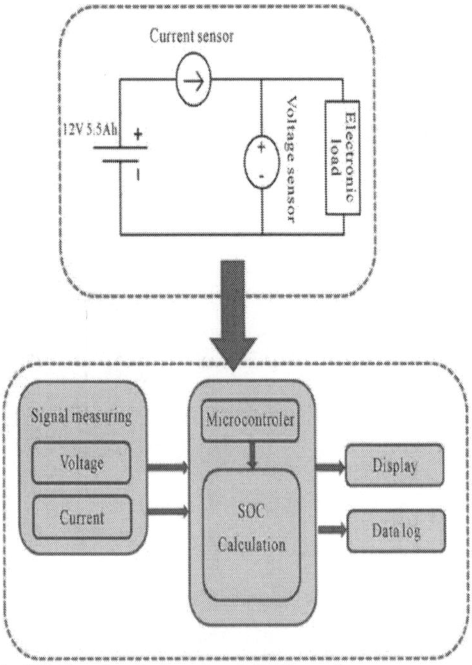

Figure 4. SOC calculation by MTB-series.

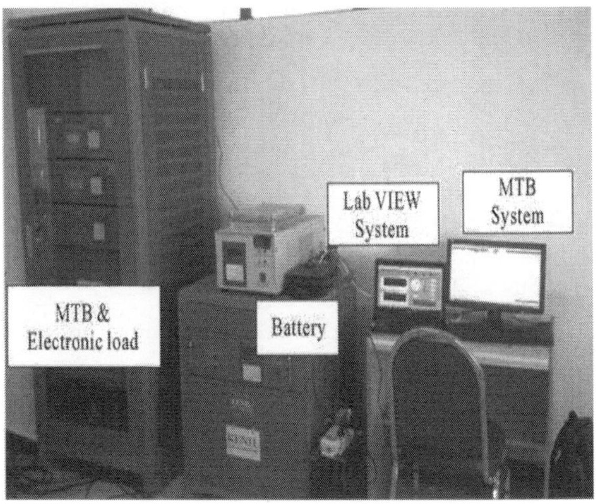

Figure 5. SOC testing system MTB-series at RMUTT Lab.

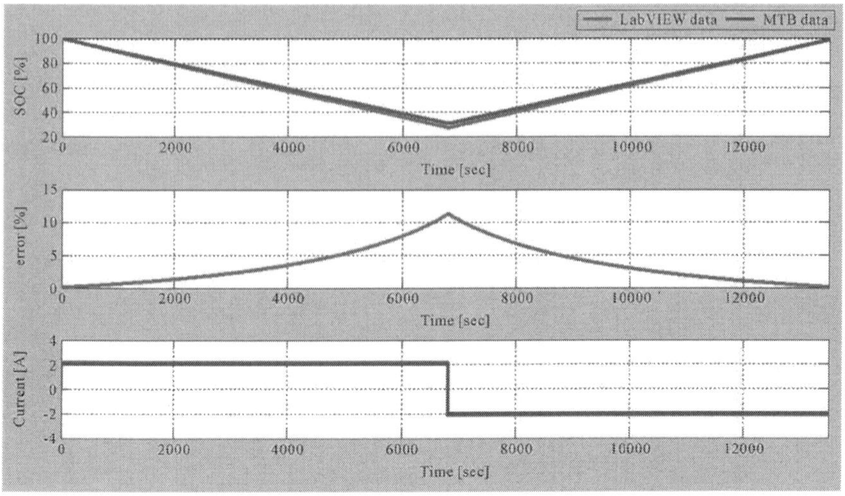

Figure 6. Results of comparison of SOC and MTB in LabVIEW.

The state of charge of the system is estimated using the Coulomb counting method in real time. The method has been verified by comparison with the Battery Testing MTB-series. The errors between the results of both systems are estimated at 5 percent. The system will issue alerts when the system detects malfunctions, and the operator can analyze and react to the system in real time to make the system stabilized and extend the lifetime of the battery. Therefore, the proposed BMS using the Coulomb counting method can be used for managing batteries in PV-hybrid systems.

REFERENCES

1. Plangklang, B. (2010) Photovoltaic System Technology. Rajamangala University of Technology, Thanyaburi.
2. http://en.wikipedia.org/wiki/Lead-acid_battery
3. Ladener, H. (1996) Solare Stromversorgung: Grundlagen, Planung, Anwenddung. ÖkobuchVerlag, Freiburg, 79-103.
4. Garche, J. and Harnisch, P. (1999) Batterien in PV-Anlagen. In: Schmid, J., Ed., Photovol-Taik: Strom aus der Sonne, Technologie, Wirtschaftlichkeit und Marktentwick-Lung, Müller, Heiderberg, 143-174.
5. Chen, Z.H. (2011) Battery State of Charge Estimation Based on a Combined Model of Extended Kalman Filter and Neural Networks. The 2011 International Joint Conference on Neural Networks (IJCNN), San Jose, 31 July-5 August 2011, 2156-2163.

6. Rodrigues, S., Munichandraiah, N. and Shukla, A.K. (2000) A Review of State-of-Charge Indication of Batteries by Means of A.C. Impedance Measurements. Journal of power Sources, 87, 12-20.
7. Pascoe, P.E. and Anbuky, A.H. (2004) VRLA Battery Discharge Reserve Time Estimation. IEEE Transaction on Power Electronic, 19, 1515.
8. Matthias, D. (2006) Dynamic Model of a Lead Acid Battery for Use in a Domestic Fuel Cell System. Journal of Power Sources, 161, 1400-1411.http://dx.doi.org/10.1016/j.jpowsour.2005.12.075
9. Olivier, T. and Louis, A.D. (2009) Experimental Validation of a Battery Dynamic Model for EV Applications. EVS24 International Battery. World Electric Vehicle Journal, 3, 1-10.
10. Shepherd, C.M. (1965) Design of Primary and Secondary Cells—Part 2. An Equation Describing Battery Discharge. Journal of Electrochemical Society, 112, 657-664.
11. Sabine, P., Marion, P. and Andresa, J. (2001) Method for State of Charge Determination and Their Application. Journal of Power Sources, 96, 113-120.
12. Sukrrmoak, K. (2009) Deteioration Factor of Battery. PEC Technology (Thailand) ltd., Bangkok.
13. IEEE (2005) Guide for Selection and Sizing of Batteries for Uninterruptible Power Supply. IEEE Std., 1184-1994.
14. Jossen, A. (1999) Battery Management Systems (BMS) for Increasing Battery Life Time. Center for Solar Energy and Hydrogen Research (ZSW Ulm), Helmholtzstr.
15. Atia, Y., Zahran, A.M. and Al-Hossain, A. (2010) Solar Cell Curves Measurement Based on LabVIEW Microcontroller Interfacing. Proceedings of the 12th WSEAS International Conference on Automatic Control Modeling & Simulation, 59-64.

CITATION

Ausswamaykin, A. and Plangklang, B. (2014) Design of Real Time Battery Management Unit for PV-Hybrid System by Application of Coulomb Counting Method. Energy and Power Engineering, 6, 186-193. doi: 10.4236/epe. 2014.67017.

CHAPTER 10

A Novel High-Performance Lekage-Tolerant, Wide Fan-In Domino Logic Circuit in Deep-Submicron Technology

Ajay Dadoria, Kavita Khare, T. K. Gupta, R. P. Singh

Electronics & Communication Engineering, MANIT, Bhopal, India

ABSTRACT

As technology shrinks in modern era the demand on high speed, low power consumption and small chip area in microprocessors is come into existence. In this paper we have presented a new class of domino circuit design for low power consumption, faster circuit speed and high performance. Due to wide fan-in domino logic, its logic gate suffer from noise sensitivity, if we improve sensitivity, sub-threshold and gate oxide leakage current dominate in evaluation network, which increases the power consumption and reduces the performance of the circuit. The proposed circuit improves the dynamic power consumption and reduces the delay which improves the speed of the circuit. Simulation is performed in BISM4 Cadence environment at 65 nm process technology, with supply voltage 1 V at 100 MHz frequency and bottleneck operating temperature of 27°C with C_L= 1 fF. From the result average power improvement by proposed circuit 1 & 2 for 8 input OR gate is 10.1%, 15.28% SFLD, 48.56%, 51.49% CKD, 55.17%, 57.71% HSD and improvement of delay is 1.10%, 12.76% SFLD, 19.13%, 28.63% CKD, 4.32%, 15.59% HSD, 19.138%, 44.25% DFD respectively.

INTRODUCTION

As technology scales down power consumption is dominant in deep sub-micron technology. Power consumption is divided into two parts static and dynamic circuit. When we talk about low power, high speed, we prefer dynamic logic instate of static CMOS circuits, by generating proper logic at output of the circuit. All these properties of the dynamic node make it high robust circuit, to generate proper logic in dynamic CMOS circuits, so to achieve high performance. Main limitations of dynamic logic occur during cascading of similar circuit during cascading of large circuit an erroneous state occurs in dynamic logic design; another problem which come into existence is charge sharing which reduces the voltage of dynamic node. So to overcome from the problem a week PMOS transistor (with a small (W/L) ratio) placed parallel to pre-charge transistor, which is feedback from the output, the output of the circuit is connected with inverter for reducing the cascading problem [1].

Domino logic circuit is also a kind of dynamic logic circuit which is used for the high speed and high performance application. Also the domino logic circuit plays a vital role where fan in are high in any circuit. Domino circuits are widely used in high performance microprocessors, register files, ALU, DSP circuits and priority encoders in content addressable memories, such as high fan-in multiplexer or comparator circuits. A basic footer less domino logic circuit is shown in Figure 1 where in pre-charge phase (clk = 0) PMOS transistor ON and charge the dynamic node, during evaluation phase (clk = 1) the dynamic node discharge when any one of the input is ON, the output of the inverter is feed back to week keeper transistor so to maintain the charge on dynamic node.

In this paper, a new domino circuit for wide fan-in applications in ultra deep sub-micrometer technologies is proposed. The novelty of the proposed circuit is that our work simultaneously reduces the delay and leakage power consumption.

The rest of the paper is organized as follows. Section 1 explains some basics of domino logic circuits. Typical conventional approach has been discussed in Section 2. Section 3 gives the brief of proposed circuit for the betterment of the noise margin. Simulation result and conclusion are discussed in Section 4 and Section 5 respectively.

LITERATURE REVIEW

The most popular domino logic is the standard footless domino [2] circuit as shown in Figure 1. For enhance the performance of the microprocessor, modification is done on circuit level to increase the robustness of the circuit, without penalty of noise immunity, the keeper transistor is added. Keeper transistor is feedback from the output; keeper transistor W/L ratio is very low to maintain the charge in the dynamic node.

The keeper ratio K is defined as

$$K = \frac{\mu p\left(\frac{W}{L}\right)\text{Keeper} - \text{transistor}}{\mu n\left(\frac{W}{L}\right)\text{evaluation} - \text{network}} \tag{1}$$

where W and L denote the transistor size, and mobility of electron and hole is represented by µn and µp. However by inserting the keeper transistor performance of the circuit is degraded and power dissipation increases. Upsizing the keeper transistor is another solution for improves robustness but it result in higher power dissipation and delay [3]. Contention between evaluation network and keeper transistor, therefore size of the keeper transistor should be low to achieve high-speed in microprocessor. Thus, trade off exist between delay and power to improve noise and leakage immunity [4]. Several techniques introduce in this section to address this issue.

High Speed Domino Logic (HS)

High speed domino is another domino logic circuit. In domino logic circuit current drawn through the keeper transistor and pull down network NMOS transistors at the beginning of the evaluation phase, can be reduced by applying a clock delay in the circuit. That does not affect the leakage current in the circuit [5]. But apart from this the extra clock delay consumes extra area and power, which is a big drawback of the circuit as shown in Figure 2. In High speed domino logic circuit when clock becomes high, M_{n1} is still off and M_{p2} is still on. Therefore M_{p2} turns off the keeper transistor. After some delay of inverter M_{p2} becomes off. Hence at the beginning of evaluation phase dynamic node is afloat, so in the absence of keeper transistor, evaluation node may be discharged for any noise at the input section [7]. Also the voltage at the gate of the keeper transistor would be V_{DD}-V_{tMn1}. This would provide a dc current flow through the PMOS keeper transistor and the NMOS network.

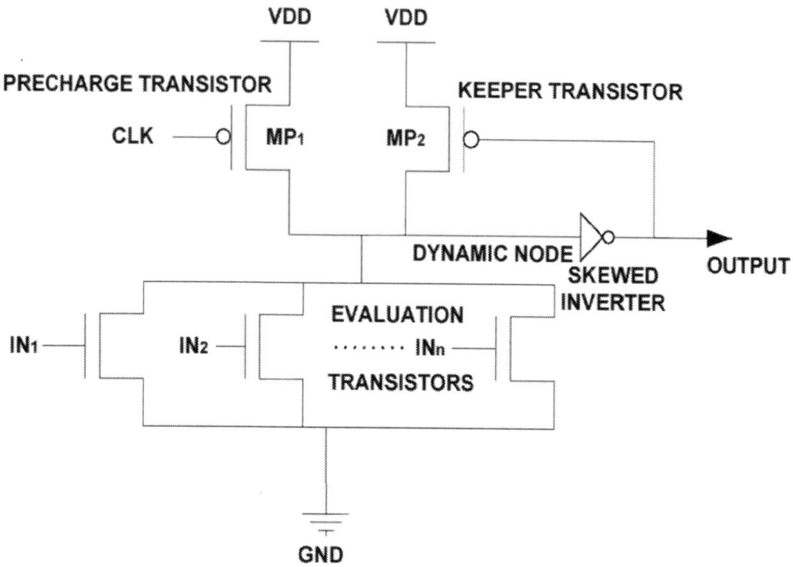

Figure 1. Standard footer less domino logic.

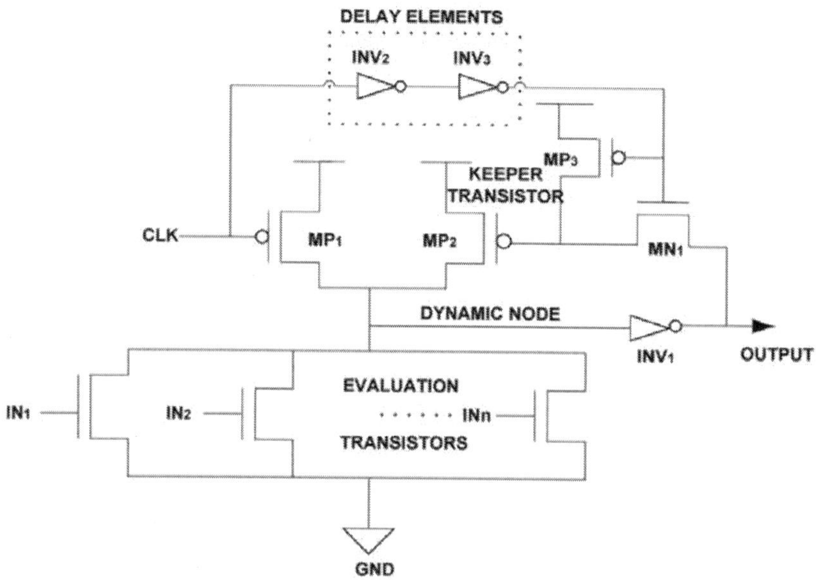

Figure 2. High speed domino logic.

Conditional Keeper Domino Logic (CKD)

The conditional keeper domino logic contains two PMOS transistor keeper circuit [6]. In which one is of smaller strength and other is of higher strength as shown in Figure 3. Now when the dynamic node is at high voltage M_{kp1} gets turn on to avoid voltage drop at the dynamic node. If the dynamic node is still high, then after a certain amount of delay, during the evaluation phase output of NAND gate becomes low which makes M_{k2} to turn on? For maintaining the state of dynamic node, is M_{k1}responsible during the beginning of evaluation phase and M_{k2} is responsible for the rest of the evaluation phase [7].

Diode Footed Domino (DFD)

In diode footed domino we modify the conventional domino circuit by adding an nMOS transistor M_1in series with the foot of the evaluation network. This nMOS transistor is in diode configuration i.e. gate and drain terminals connected together. Figure 4 [9] shows the Diode Footed Domino configuration. Stacking effect [10] occur because this transistor M_1 is connected in series with the evaluation network [12] [13] . Thus sub-threshold leakage current reduces as a result of stacking effect. DFD circuit works as follow:

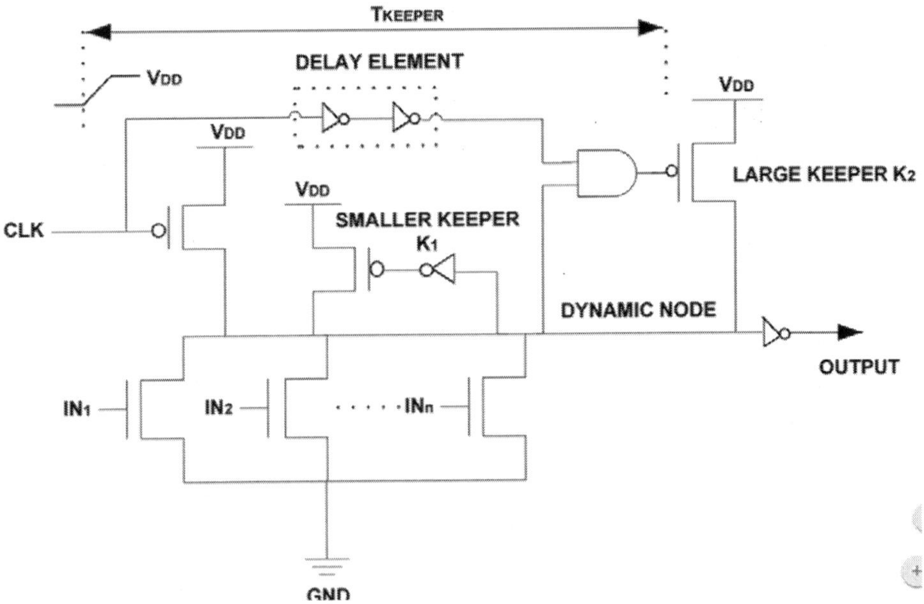

Figure 3. Conditional keeper domino logic.

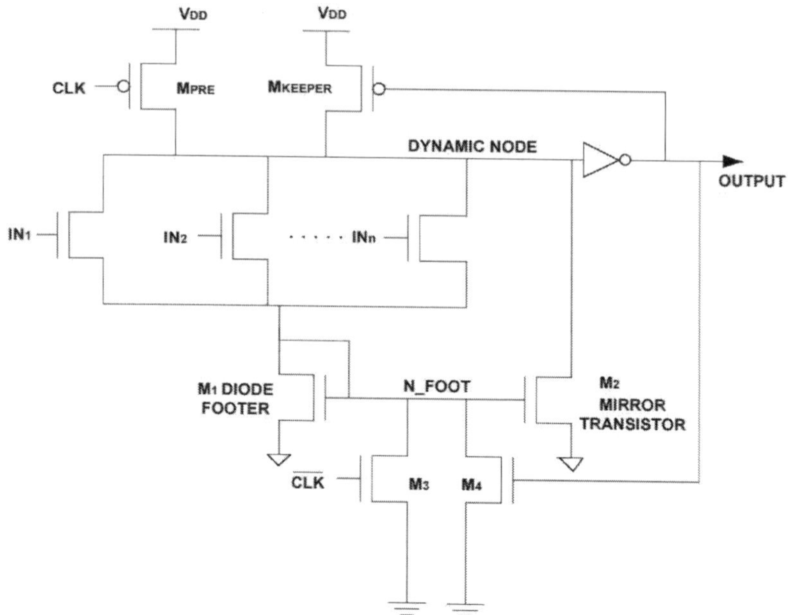

Figure 4. Diode footed domino logic.

The mirror ratio is defined as the ratio of the current drivability of the mirror transistor to that of the diode footer:

$$M = \frac{\left(\frac{W}{L}\right)_{\text{mirror transistor}}}{\left(\frac{W}{L}\right)_{\text{diode footer}}}$$

By increasing the mirror ratio, the performance can be increased.

PROPOSED CIRCUIT

In conventional footer less circuit, during precharge phase (clk = 0), PMOS transistor ON and charge the dynamic node from V_{dd}, during evaluation phase (clk = 1), a dynamic node not able to maintain the constant because PMOS transistor rail OFF from $V_{dd,}$ only keeper Transistor connected to V_{dd}maintain the charge of dynamic node if all the transistor is OFF in evaluation network as shown inFigure 5. During evaluation phase when any one input is ON of NMOS block the dynamic node will discharge, which result in flow of sub-thre- shold and gate oxide leakage current which result in degradation of UNG of the circuit, for reduction of leakage current and enhance the noise immunity of the circuit we have proposed a circuit.

In proposed circuit modification is done in evaluation network, we have inserted two NMOS transistor between dynamic node and pull down network. To improve the efficiency of the proposed circuit and extra NMOS transistor is connected to the dynamic node to produce the proper stacking of the evaluation network, to increase the noise immunity of the circuit and reduces the leakage current of the circuit by providing half swing logic at the output node as sown in Figure 6. In footed portion we place NMOS transistor, during pre-charge phase footed transistor is OFF, during evaluation phase a charge discharge from dynamic node the two NMOS transistor provides the stacking effect for leakage reduction and high noise immunity. In proposed circuit 2 as shown in Figure 7. Input of footed transistor is connected to clock and three inverter for generation of delay in the footed transistor, the delay element is used for proper slower the gate and greater noise robustness. These approaches do not reduce the overall leakage current, but only the leakage current at the dynamic node that drives the final static inverter and is the critical node. Hence we have more degree of freedom for increasing speed or enhance noise immunity by reducing the leakage current.

SIMULATION RESULTS AND COMPARISON

The proposed circuit was simulated using BISM4 Cadence virtuoso schematic editor tool in the high-perfor- mance 65-nm predictive technology and at the temperature of 27°C. The supply voltage used in the simulations is 1 V, for 8, 16 and 32 input OR gates as shown in Tables 1-4. Various parameters have been mitigated such as power dissipation, delay and UNG (Unit Noise Gain). Table 4 calculate the standby leakage current at different temperature for 8, 16 and 32 inputs, operating frequency is 100 MHz with $C_L = 1$ pF of the circuit. Average power consumption of existing domino logic circuit and proposed circuit is shown in Figure 8. UNG graph is shown in Figure 9.

The noise-margin metric used in this paper is called the unity noise gain (UNG). For calculation of UNG [11], a pulse noise is applied to all inputs with amplitude which is a fraction of supply voltage and a pulse width equal to 30% of duty cycle. This noise amplitude is defined as

$$\text{UNG} = \{\text{Vin}, \text{Vnoise} = \text{Voutput}\}$$

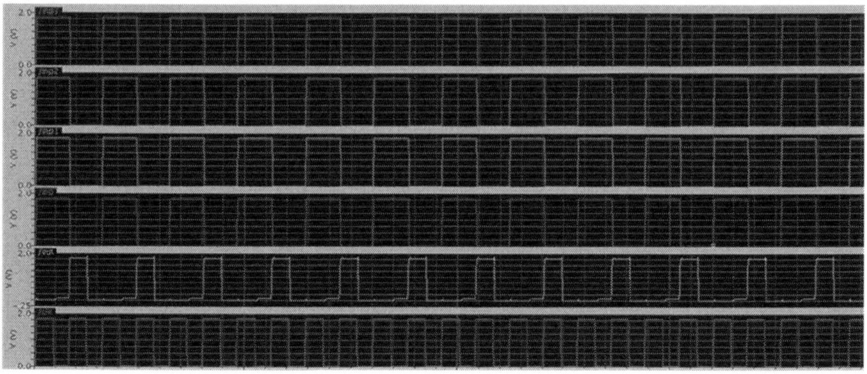

Figure 5. Output wave form of proposed circuit.

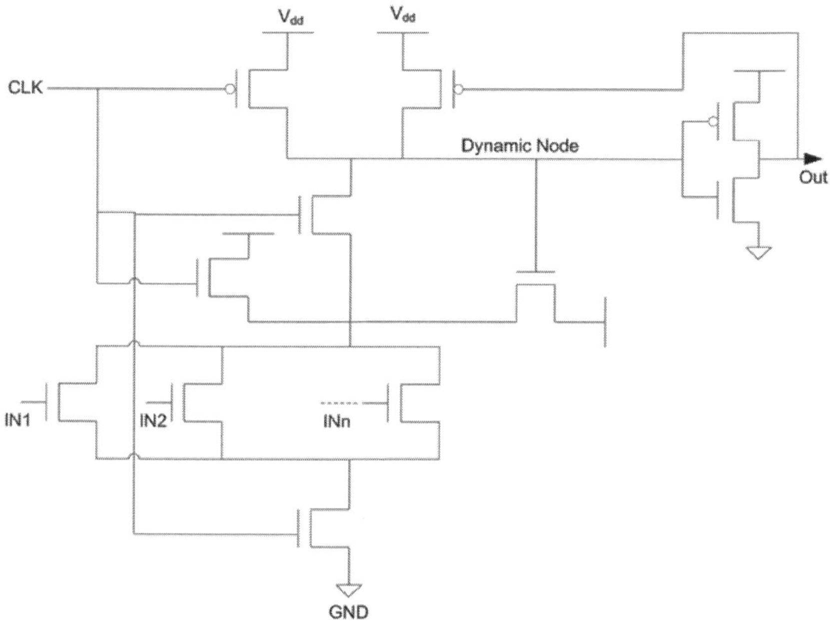

Figure 6. Proposed circuit 1.

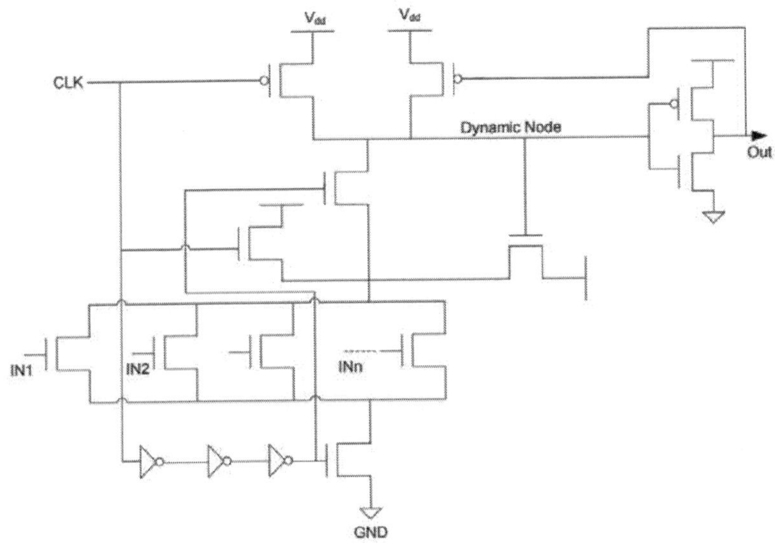

Figure 7. Proposed circuit 2.

Table 1. Comparison of power dissipation (μW).

S. No.	Logic Style	8 Input	16 Input	32 Input
1.	SFLD	2.611	4.234	7.632
2.	CKD	4.56	6.323	9.862
3.	HSD	5.231	8.214	10.12
4.	DFD	1.672	2.540	3.965
5.	Proposed Circuit 1	2.345	5.203	8.812
6.	Proposed Circuit 2	2.212	4.780	7.642

Table 2. Comparison of UNG (in Volt).

S. No.	Logic Style	8 Input	16 Input	32 Input
1.	SFLD	0.252	0.228	0.212
2.	CKD	0.264	0.225	0.199
3.	HSD	0.243	0.230	0.203
4.	DFD	0.274	0.256	0.2574
5.	Proposed Circuit 1	0.263	0.231	0.203
6.	Proposed Circuit 2	0.254	0.224	0.191

Table 3. Comparison of Delay (in ps).

S. No.	Logic Style	8 Input	16 Input	32 Input
1.	SFLD	13.102	18.435	28.59
2.	CKD	16.015	21.2	31.23
3.	HSD	13.541	18.32	28.43
4.	DFD	16.49	22.16	33.12
5.	Proposed Circuit 1	12.95	15.651	21.34
6.	Proposed Circuit 2	11.432	14.125	19.57

Table 4. Stand by leakage current of proposed circuit (in nA).

S. No.	Temperature	8 Input	16 Input	32 Input
1.	27°C	262.25	308.79	401.11
2.	50°C	364.12	411.34	505.45
3.	110°C	700.24	754.23	851.34

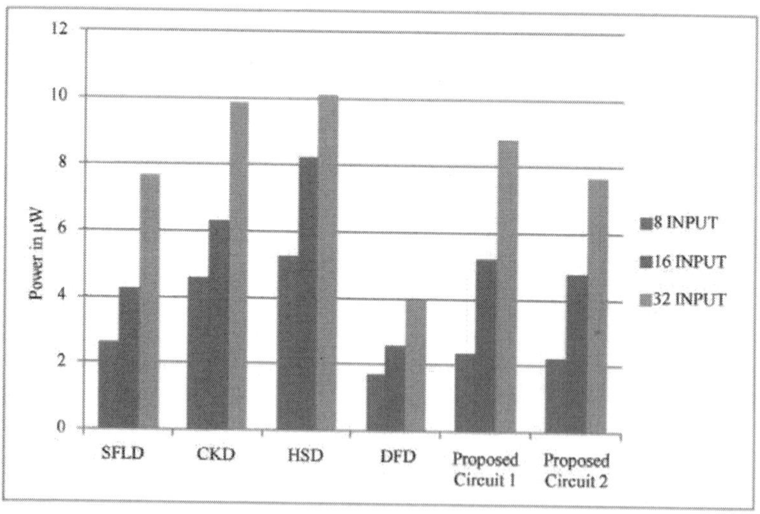

Figure 8. Comparison of average power.

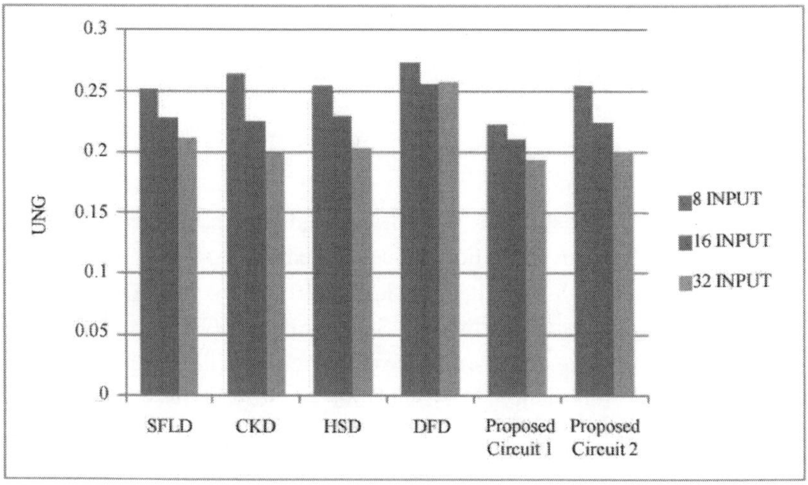

Figure 9. Comparison of UNG.

CONCLUSION

Domino logic dissipates very low standby power compare to static CMOS logic. The novelty of the proposed circuit is that domino circuit reduces the leakage power consumption by maintaining the same level of delay and UNG. In this paper our main objective was to improve the noise immunity and to reduce the average power consumed and delay associated with the circuit. The analysis shows that we get an appreciable improvement in the noise margin and power consumption and marginal improvement in delay. The whole comparison is based upon 65 nm CMOS technology using Cadence Virtuoso tool.

REFERENCES

1. Clarke, L.T. and Taylor, G.F. (1996) High Fan-In Circuit Design. IEEE Journal of Solid-State Circuits, 31, 91-96. http://dx.doi.org/10.1109/4.485870
2. Tsui, B.-Y. and Chin, L.-F. (2004) A Comprehensive Study of the FIBL of Nanoscale MOSFETs. IEEE Transactions on Electron Devices, 51, 1733-1735.
3. Moradi, F., VuCao, T., Vatajelu, E.I., Peiravi, A., Mahmoodi, H. and Wisland, D.T. (2012) Domino Logic Designs for High-Performance and Leakage-Tolerant Applications. Elsevier INTEGRATION, the VLSI Journal, 46, 247-254.

4. Moradi, F., Peiravi, A. and Mahmoodi, H. (2004) A New Leakage Tolerant Design for High Fan-in Domino Circuits. IEEE Conference, 493-496.
5. Allam, M.W., Anis, M.H. and Elmasry, M.I. (2000) High-Speed Dynamic Logic Style for Scaled-Down CMOS and MTCMOS Technologies. Proceedings of the International Symposium on Low Power Electronics and Design, 155-160.http://dx.doi.org/10.1109/LPE.2000.876774
6. Alvandpour, A., Krishnamurthy, R.K., Soumyanath, K. and Borkar, S.Y. (2002) A Sub-130-nm Conditional Keeper Technique. IEEE Journal of Solid-State Circuits, 37, 633-638.http://dx.doi.org/10.1109/4.997857
7. Alvandpour, A., Krishnamurthy, R., Soumayanath, K. and Borkar, S. (2001) A Low-Leakage Dynamic Multi Ported Register File in 0.13 μm CMOS. Proceedings of International Symposium on Low Power Electronics and Design, 68- 71.
8. Peiravi, A. and Elissa, M.A. (2013) Current-Comparison-Based Domino: New Low-Leakage High-Speed Domino Circuit for Wide Fan-In Gates. IEEE Transactions on Very Large Scale Integration (VLSI) Systems, 21.
9. Lih, Y., Tzartzanis, N. and Walker, W.W. (2007) A Leakage Current Replica Keeper for Dynamic Circuits. IEEE Journal of Solid-State Circuits, 42, 48-55.http://dx.doi.org/10.1109/JSSC.2006.885051
10. Yorozu, Y., Hirano, M., Oka, K. and Tagawa, Y. (1987) Electron Spectroscopy Studies on Magneto-Optical Media and Plastic Substrate Interface. IEEE Translation Journal on Magnetics in Japan, 2, 740-741. [Digests 9th Annual Conf. Magnetics Japan, p. 301, 1982].
11. Peiravi, A., Moradi, F. and Wisland, D.T. (2009) Leakage Tolerant, Noise Immune Domino Logic for Circuit Design in the Ultra Deep Submicron CMOS Technology for High Fan-In Gates. Journal of Applied Sciences, 9, 393-396.http://dx.doi.org/10.3923/jas.2009.392.396
12. Mahmoodi-Meimand, H. and Roy, K. (2004) Diode-Footed Domino: A Leakage-Tolerant High Fan-in Dynamic Circuit Design Style. IEEE Transactions on Circuits and Systems—I, 51, 495-503.
13. Gupta, T.K. and Khare, K. (2014) A New Technique for Leakage Reduction in 65 nm Footerless Domino Circuits. Scientific Research Circuits and Systems, 4

CITATION

Dadoria, A., Khare, K., Gupta, T. and Singh, R. (2015) A Novel High-Performance Lekage-Tolerant, Wide Fan-In Domino Logic Circuit in Deep-Submicron Technology. *Circuits and Systems*, **6**, 103-111. doi:10.4236/cs.2015.64011.

CHAPTER 11

Analog Encoding Voltage—A Key to Ultra-Wide Dynamic Range and Low Power CMOS Image Sensor

Arthur Spivak [1], Alexander Belenky [1], Alexander Fish[2] and Orly Yadid-Pecht[3]

[1]The VLSI Systems Center, LPCAS, Ben-Gurion University, P.O.B. 653, Be'er-Sheva 84105, Israel
[2]Department of Electrical and Computer Engineering, Bar Ilan University, Ramat Gan, 52100, Israel
[3]Department of Electrical Engineering, University of Calgary, Alberta T2N 1N4, Canada

ABSTRACT

Usually Wide Dynamic Range (WDR) sensors that autonomously adjust their integration time to fit intra-scene illumination levels use a separate digital memory unit. This memory contains the data needed for the dynamic range. Motivated by the demands for low power and chip area reduction, we propose a different implementation of the aforementioned WDR algorithm by replacing the external digital memory with an analog in-pixel memory. This memory holds the effective integration time represented by analog encoding voltage (*AEV*). In addition, we present a "ranging" scheme of configuring the pixel integration time in which the effective integration time is configured at the first half of the frame. This enables a substantial simplification of the pixel control during the rest of the frame and thus allows for a significantly more remarkable DR extension. Furthermore, we present the implementation of "ranging" and *AEV* concepts on two different designs, which are targeted to reach five and eight decades of DR, respectively. We describe in detail the operation of both systems and provide the post-layout simulation results for the second solution. The simulations show that the second design reaches DR up to 170 dBs. We also provide a comparative analysis in terms of the number of operations per pixel required by our solution and by other widespread WDR algorithms. Based on the calculated results, we conclude that the proposed two designs, using "ranging" and *AEV* concepts, are attractive, since they obtain a wide dynamic range at high operation speed and low power consumption.

INTRODUCTION

Extending the dynamic range (DR) of a CMOS image sensor's (CIS) remains one of the challenges to be faced in designing an effective versatile sensor. Improvement of image capture capability can be done either by reducing the noise floor (NF) of the sensor [1, 2, 3] or by extending its saturation toward higher light intensities. Most of the cases that are reported in the literature give solutions which focus on extending the DR toward high light intensities. There are numerous solutions that have been proposed throughout the years [4,5,6,7,8,9,10,11,12,13,14,15,16,17,18,19,20,21,22,23,24,25,26,27].

However, the majority of the proposed WDR solutions have used an external memory, which significantly increased their area and power consumption and, as a result, their cost.

Examples of using a digital memory can be found in a large variety of solutions [8] such as (a) multimode sensors that have a linear and a logarithmic response at dark and bright illumination levels, respectively [9]; (b) clipping sensors, in which a well capacity adjustment method is applied [10]; (c) frequency-based sensors, in which the sensor output is converted into a pulse frequency [12]; (d) time-to-first spike (TFS) sensors, in which the image is processed according to the time the pixel was detected as saturated [9]; (e) sensors with global control over the integration time, in which the pixel integrates for a set of exposures regardless of its input light intensity [15]; and (f) sensors with autonomous control over the integration time (multiple resets), in which each pixel has control over its own exposure period [16,17,18]. In the aforementioned designs the memories were used to store the possible pixel outputs or the DR extension data. The reason for using a memory is that the data processing becomes more straightforward, *i.e.*, it is easy to split the processing into several independent stages and to store the intermediate results. However, in such a case, the overall processing time and power increase.

Nevertheless, we can find memory free WDR solutions as well. For example, in [19] a TFS algorithm implementation is described, where, instead of using a digital code to represent the saturation time, the authors paired each time slot with a certain voltage value that was subsequently converted into corresponding digital pixel value. This solution unfortunately suffers from the reduced sensitivity throughout the whole DR [8].

It is especially interesting to look into the free memory solutions in [20, 21], which are based upon the global control over the integration time, since this category is the most widely spread in the industry. The memory can be omitted if there are two exposures as will be shown later on. However, using two exposures causes a substantial signal to noise ratio (SNR) drop in the mid-tones, which can necessitate subsequent image processing.

With autonomous control over the integration time category, the memory emerges as the main cause for the increase in both the complexity and the cost of the solution. Attempts to dump the memory in this category have been undertaken as well. In [22] a successful memory free solution was presented. That solution was based upon a very simple pixel structure, in which the pixel integration time value was encoded in voltage and stored on the intra-pixel capacitance. But, since the same capacitance was also used for readout of the intermediate pixel values, the stored integration times were continuously overwritten and refreshed over and over again slowing the sensor operation.

In this work, we present two novel concepts in "ranging" and analog encoding voltage (*AEV*) for design of low power, high speed, memory free WDR sensors with autonomous control over integration time. According to the "ranging" concept, the pixel signal is "coarsely" quantized at the beginning of the frame, and the final "fine" quantization takes place at the end of the frame. Generated data in the "coarse" quantization is stored as analog encoding voltage (*AEV*) inside each pixel on a separate capacitance, and this controls the pixel integration throughout the frame. Since we use different capacitances for a photo-induced signal and for an (*AEV*), the data processing becomes very simple and fluent. At the end of the frame, both of the two analog values, the photo-generated charge and the *AEV*, are merged through an analog to digital A/D conversion to a final digitized pixel value.

In addition, we present two designs which implement the "ranging" and *AEV* concepts. The designs are thoroughly described from the schematic to the layout. Moreover, we present the post-layout simulation results of the second design, showing successful DR extension up to 170 dBs. To illustrate the effectiveness of the proposed solutions, we perform a qualitative comparison with other published solutions and show that eventually our solutions are optimized for low power WDR imaging.

The presented work is organized as follows: Section 2 introduces our solutions, Section 3 presents the comparison of a number of operations in several types of WDR solutions, and Section 4 concludes the study. Following the last section, there are two Appendixes, namely A and B. They include the detailed considerations of pixel swing segmentation and signal to noise (SNR) considerations, respectively. In this way, the overall concept becomes clearer and no important design consideration is omitted.

WDR SOLUTION BASED UPON RANGING AND AEV

Problem Definition and Proposed Solution

The solution we have used for the DR extension is called the autonomous control over the integration time (multiple resets). None of the pixels in this solution are classic 3T's, because this enables individual pixel reset via an additional transistor [8]. During the frame the outputs of a selected row are read nondestructively through the regular output chain. Then they are compared with an appropriate threshold at certain time points by comparators, which are found at a separate unit outside the pixel array. If a pixel value exceeds the threshold, a reset is given at that time point to that pixel. The binary information concerning the reset being applied or not is saved in a digital storage, in order to enable proper scaling of the value read. This enables the pixel value to be described as a floating-point representation, wherein the exponent will describe the scaling factor for the actual integration time, while the mantissa will be the regular A/D output. In this way the actual pixel value would be:

$$V_{pixel} = M_{pixel} X^{EXP} \quad (1)$$

where V_{pixel} is the actual pixel value, M_{pixel} is the analog or digitized output value that has been read out at the end of the frame, X is a chosen constant, 2, for instance. EXP is the exponent value, which describes the scaling factor, *i.e.*, the specific part of the integration time the pixel actually integrated without being saturated.

The bottleneck of the basic solution is that bits representing the EXP are generated during the integration sequentially; hence, after the generation of the subsequent bit, a digital memory should be refreshed. As the number of pixel rows increase, the time that can be allocated to EXP bits' generation becomes more limited. The restrictions on the DR extension vary according to the mode of sensor operation: rolling or global shutter. This is more obvious in the rolling shutter, wherein the integration in the successive rows is skewed, so there is more time for DR processing. On the other hand, in the global shutter sensor, wherein the integration starts simultaneously in all rows, the processing times are substantially reduced. Therefore, the solution that we propose is targeted mainly for the global shutter sensor.

The Proposed Solution

Our solution is based upon "ranging" and *AEV* concepts. The purpose of this solution is to reduce DR processing times and to omit the memory unit. In addition, we have presented the solution architecture and the flow.

We have divided the WDR algorithm into three phases (Figure 1). During Phase 1, the *EXP* values are produced and memorized inside each pixel in a format of analog encoding voltage *AEV*. This assignment, which precedes the saturation checks, is called "ranging", since it "coarsely" quantizes [23] the pixel signal. In fact, the *AEV* value produced in this quantization, contains the most significant bits of pixel signal $Pix<M + L - 1:L>$ (Figure 2); whereas the *Mantissa* obtained at the end of the next phase contains the least significant bits $Pix<L - 1:0>$. In this way, each pixel is given its valid integration time for the current frame before the saturation checks start.

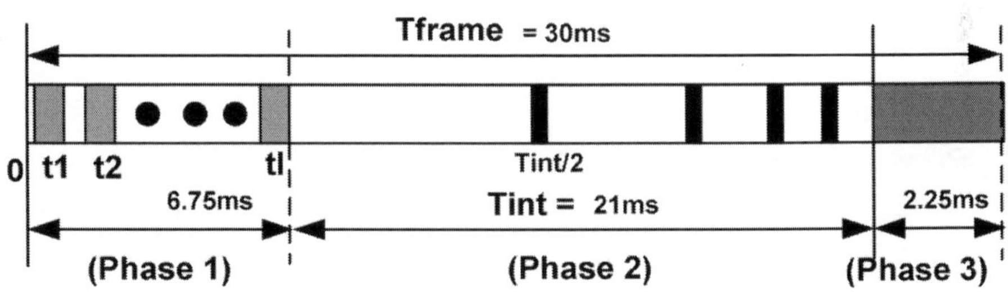

Figure 1. Three Phases of the proposed WDR algorithm.

During Phase 2, the pixel integrates in accordance with its *AEV*. Since the integration time is already known, there is no need to retrieve the last reset information or to refresh information for the current check as was performed in our previous solutions; we merely compare the *AEV* to the global reference. In case the *AEV* is below the reference, the pixel will be reset; if not, the pixel will remain untouched till the next frame.

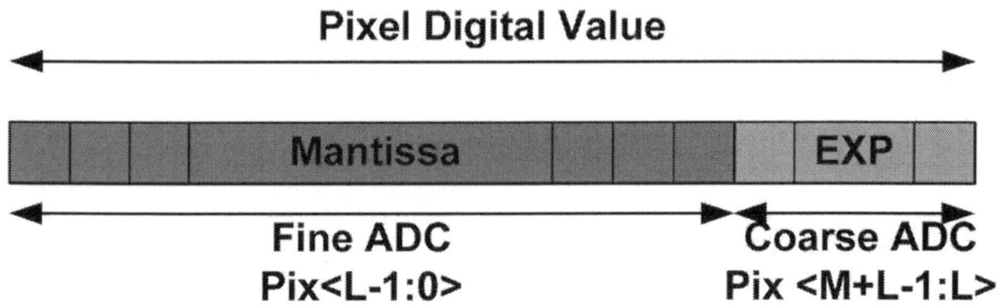

Figure 2. Digital Pixel Value: Mantissa and the EXP bits.

During Phase 3, both the *AEV* and *Mantissa* are digitized in a single slope analog to a digital conversion (ADC) (Figure 3). For this purpose, the *AEV* and *Mantissa* are read separately and converted with different resolutions. The *AEV* is digitized with a "coarse" resolution; whereas the *Mantissa* is converted by a ramp with a "fine" resolution.

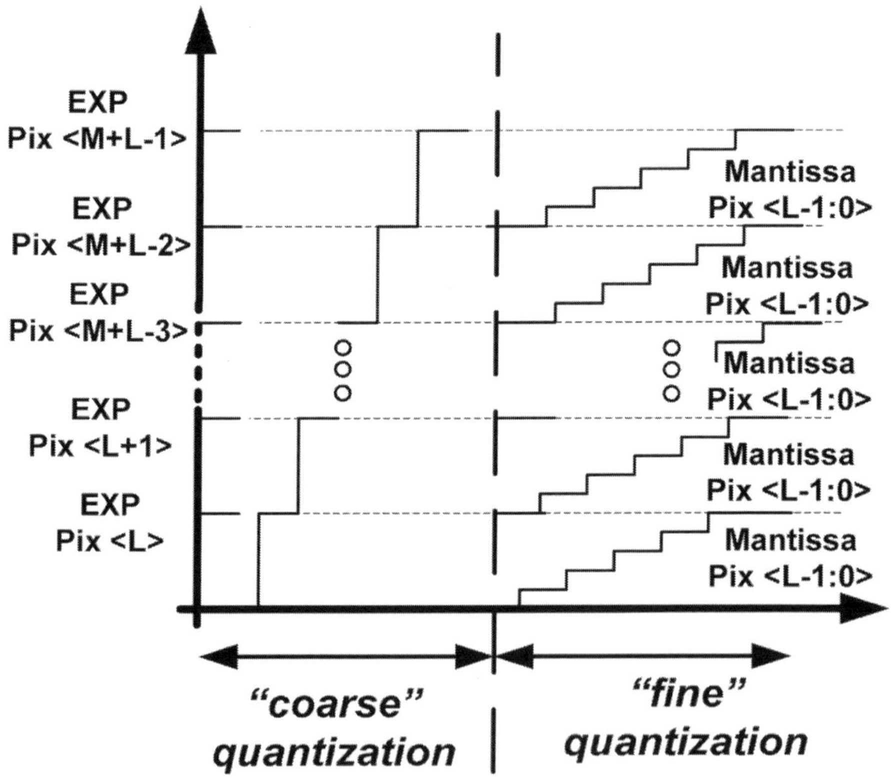

Figure 3. The Course and Fine Quantization of the Pixel Signal.

In general, the three phases can overlap each other. However, in this study, we present the first version of the algorithm; therefore, we assume that each phase is completely separated from the others. In future works we will demonstrate how these phases can be overlaid on each other. We assessed the duration of each phase in accordance with further presented simulation results. In Figure 1, we present the division of a single frame (30 ms) into three phases as if the algorithm is applied on a 1000 × 1000 pixel array. In spite of all the phases being separate, it is important to note that the maximal integration time (Phase 2) still occupies the major part of the frame, whereas the overall integration times in the rest of the phases 1 and 3 are substantially shorter.

Phase 1: Assigning the Analog Encoding Voltage (AEV)

How are the values of *AEV* selected and paired with the corresponding *EXP* bits? The *AEV* values can be dispersed within the voltage domain in several ways: logarithmically or linearly (Figure 4). In this study we chose, for the sake of convenience, a linear quantization, according to which the all the *AEV*values are spaced at the same distance of $2\Delta V_{min}$. We also decided that the *AEV* corresponding to the highest light intensity would be represented by the lowest available voltage, since this simplifies all the subsequent conversions. The parameter ΔV_{min} is derived from the properties of the comparator that are to be employed for assigning the *AEV* values. The requirements for ΔV_{min} are intuitive and straightforward:

$$\Delta V_{min} \geq \frac{V_{out\_comparator}}{G}, \quad \Delta V_{min} \geq V_{offset} \tag{2}$$

in which $V_{out\_comparator}$, G, V_{offset} are the comparator's output swing, gain, and offset, respectively.

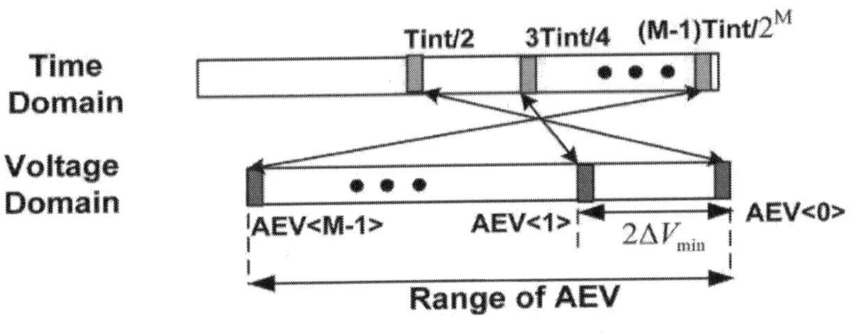

Figure 4. Analog encoding voltage (*AEV*) concept: each dynamic range (DR) bit has its own Analog Encoding Voltage.

After the pixel array has been reset globally, Phase 1 begins. The pixels are exposed to light and accumulate the photo-generated charge. At certain predetermined time points, each pixel row is sampled nondestructively and spanned by a piecewise increasing ramp. In case the pixel signal intersects with the ramp, the comparator flips and the corresponding *AEV* is written into the in-pixel dynamic memory. Detailed considerations of the *AEV* assignment are provided in Appendix A.

Phase 2 Performing the Conditional Resets

Phase 2 performs the conditional resets to the pixel array. Since the base of the DR extension is 2, the integration times are ordered in exactly the same way as in our previous solutions [24], *i.e.*, in geometric descending sequence:

$$\frac{T_{int}}{2}, \frac{T_{int}}{2^2} \cdots \frac{T_{int}}{2^{EXP}} \tag{3}$$

Nonetheless, in the current solution, the pixel reset cycle can be reduced by a factor of 2 at least, since the number of operations before the conditional reset is halved. All that has to be done is to compare the individual *AEV* with the reference, which corresponds to the current reset cycle. This phase can be performed either in a traditional fashion, *i.e.*, row by row scan, or instantly within the whole array. We will discuss in further detail two different designs implementing both of the options, namely **TYPE I** and **TYPE II.**

Phase 3: Coarse, Fine ADC & Readout

Phase 3 concludes the frame. At this stage, the analog pixel value (*Mantissa*) and the *AEV* are converted by the same comparators, which had been used in the two preceding phases. In fact, we performed a single slope ADC to obtain the digitized values of *Mantissa* and *AEV* (Figure 2). The conversion is performed row by row: first, the *AEV* is spanned with the "coarse" resolution, and then the pixel swing is spanned by the ramp with "fine" resolution. Of course, after the span, both the *AEV* and pixel capacitances are reset to their initial predetermined values.

TYPE I Pixel

For most applications the required dynamic range (DR) does not exceed 120 dBs, so the pixel structure can be straightforward. In our case, all we need is to accommodate the *AEV* and the pixel *Mantissa* on two separate capacitors and to add a readout chain and conditional access circuitry (Figure 5) to each of them. Since the pixel structure here is not complex and the DR extension is typical, this design is named **TYPE I.**

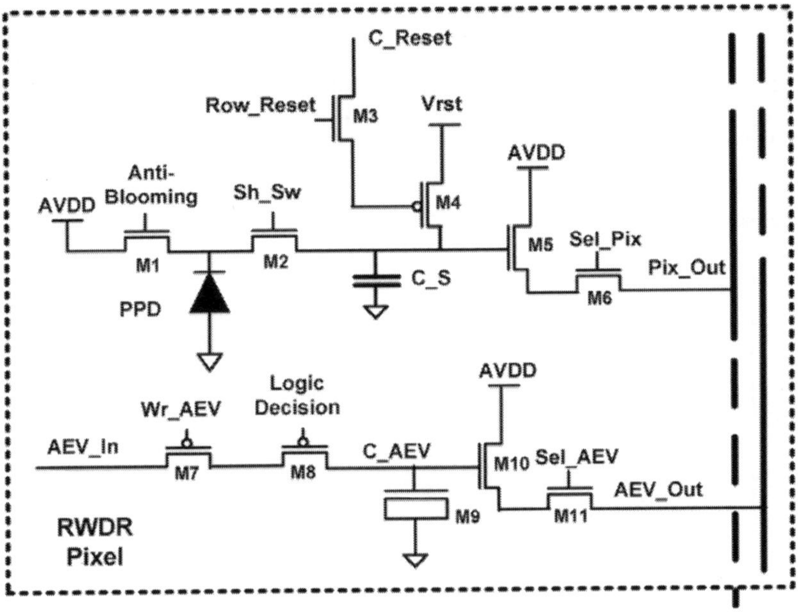

Figure 5. Schematic of **TYPE I** Pixel.

The functionality of the **TYPE I** pixel is implemented with 11 transistors only. They can be divided into two groups: the first processes the data received from the pinned photodiode (*PPD*) M_1–M_6; and the second is responsible for the *AEV* handling M_7–M_{11}.

The charge transfer from the *PPD* is bidirectional, similar to [25]. In this way, we can deliberately dump the generated charge to drain the photo-diode through M_1 or we can transfer the charge for further processing to C_s through M_2. We use a simple conditional reset scheme, implemented by M_3 and M_4transistors, to implement the multiple resets algorithm. By activating the *Row_Reset* signal, the required pixel row is chosen and then, by means of the *C_Reset* signal, driven by the column-wise comparators, is conditionally reset to V_{rst}. Transistors M_5, M_6 form a traditional source follower to handle the C_S readout [26].

AEV processing inside the pixel is very simple as well. The *AEV* value is fed to the pixel from the column-wise bus *AEV_In*. The write operation of the input *AEV* into the pixel dynamic memory is conditional and is implemented by using a trivial stacked scheme M_7 and M_8. By activating the *Wr_AEV*signal, a required row is accessed; while the *Logic Decision* signal, driven by column-wise comparators, enables the MOS capacitor C_{AEV} to sample the *AEV_In* bus. It is important to note that aside from its simplicity, this stacked scheme also minimizes the possible leakage from the *AEV* capacitor, thus maintaining the data integrity till the readout. The readout of the *AEV* is performed through a separate source follower, implemented by M_{10} and M_{11}. This relatively simple pixel was implemented in layout using 0.18 μm CMOS technology, with 14 μm pitch and 40% fill factor (FF) (Figure 6).

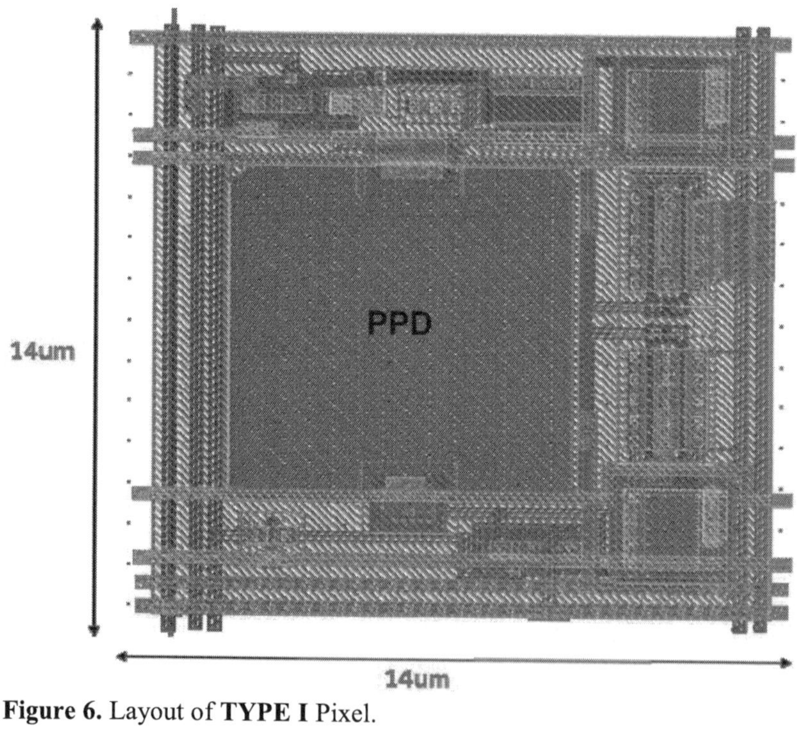

Figure 6. Layout of **TYPE I** Pixel.

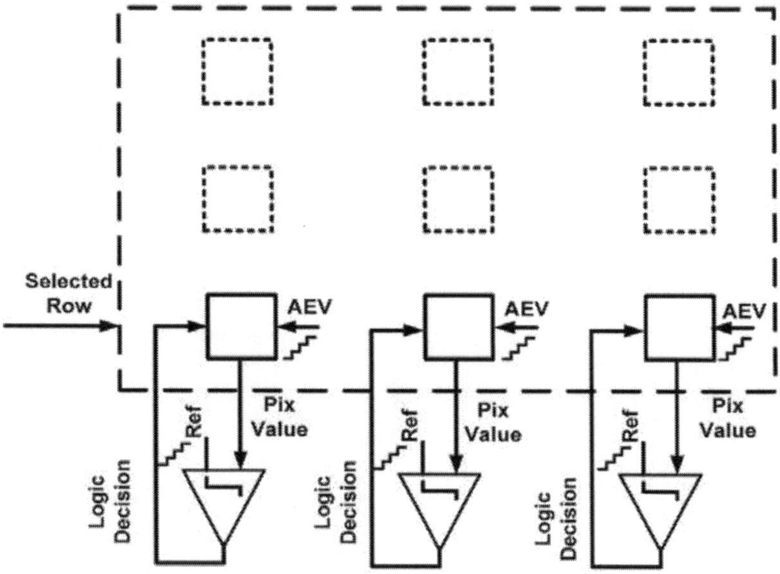

Figure 7. Phase 1 **TYPE I** pixel.

Figure 7 depicts Phase 1 for the **TYPE I** pixel. At each span, the pixel (*Pix value*) is being accessed by election of a specific row and is being compared to the reference signal *Ref.*

After Phase 1 is completed, Phase 2 begins. During this phase, each pixel integrates in accordance to the assigned *AEV*. After a certain integration slot (3) elapses, the *AEV'*s within each row are selected and compared with the *Ref*, which now reflects the reference values of integration times (Figure 8).

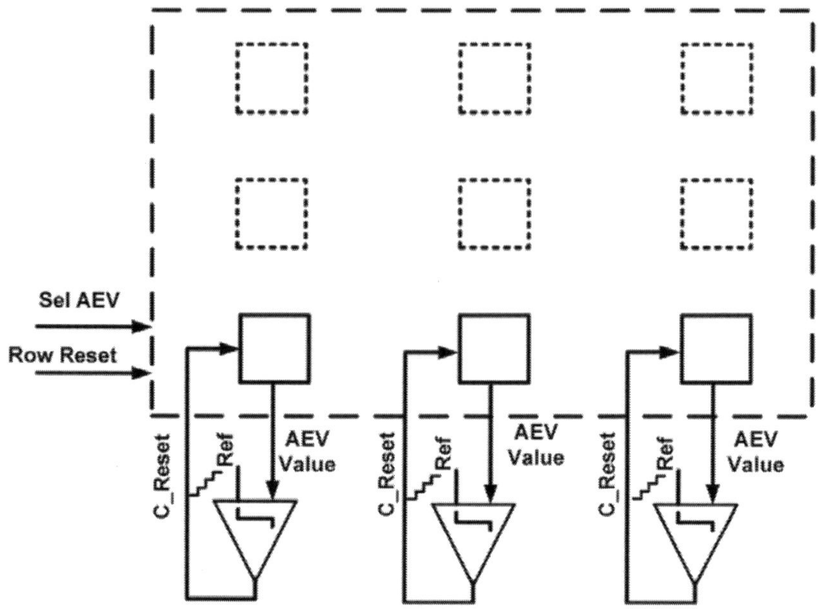

Figure 8. Phase 2 **TYPE I** pixel.

In our solution, the pixel is reset as long as its *AEV* is lower than the reference value.

In a **TYPE I** solution, the conditional resets are applied to each pixel row sequentially. The processing time, which takes to apply the conditional reset a single pixel row T_{row_reset}, is given by:

$$T_{row\_reset} = T_{AEV\_read} + T_{compare} \tag{4}$$

where T_{AEV_read} and $T_{compare}$ are the time to sample the *AEV* and to decide to reset or not the referred pixel, respectively. Since two saturation checks cannot overlap each other, the time it takes to complete a single one sets the minimal integration time $T_{\min\_\text{TYPE I}}$:

$$T_{\min\_\text{TYPE I}} = N_{rows} T_{row\_reset} \tag{5}$$

Given the minimal integration time, we can easily calculate the maximal possible DR extension factor (DRF) [8]. Through the performed simulations, we learned that the minimal integration time $T_{\min\_\text{TYPE I}}$ is 112 μs. Assuming the maximal integration time is 30 ms, we have obtained a DR extension factor equal to 256 (48 dB). The expected intrinsic DR *i.e.*, before the extension is 60 dBs, consequently the total DR equals 108 dBs.

In the last phase of the frame (Figure 9), the accumulated data are accessed for the final ADC. By activating two separate select signals *Sel_Pix* and *Sel_AEV*, the pixel and the *AEV*, respectively, are digitized using the same comparators as in Phase 1

The **TYPE I** design was successfully tested in various post-layout simulations, which proved its feasibility. To reduce the length of this work, we present only the post-layout simulation results of the**TYPE II** design because it is much more complex. To conclude, CMOS sensor based upon the presented**TYPE I** pixel is capable of providing a DR of almost 5 decades operating at video frame rate and providing an image with a decent resolution.

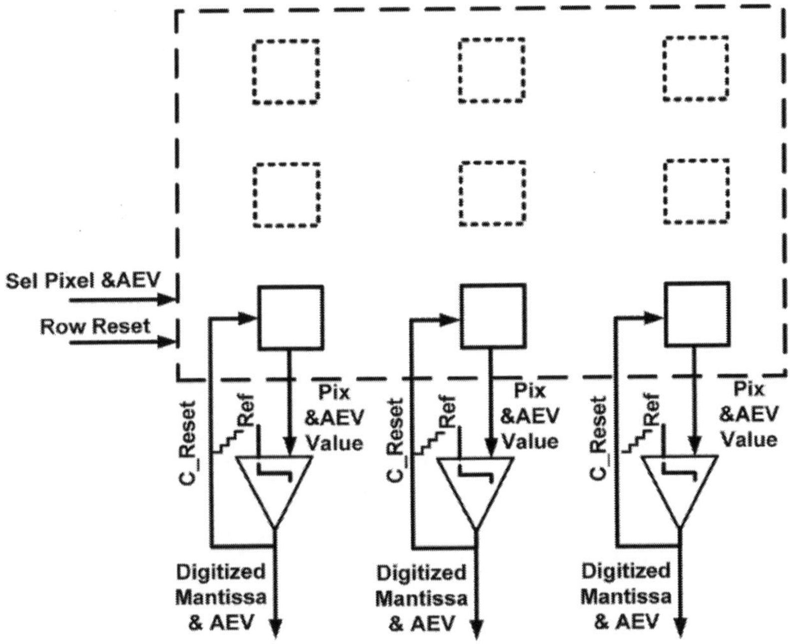

Figure 9. Phase 3 **TYPE I** pixel.

Ultra WDR (TYPE II) Pixel

The main difference between the Ultra WDR (**TYPE II**) pixel and the previous **TYPE I** solution lies in its accommodating a differential stage (*D*) within the pixel (Figure 10).

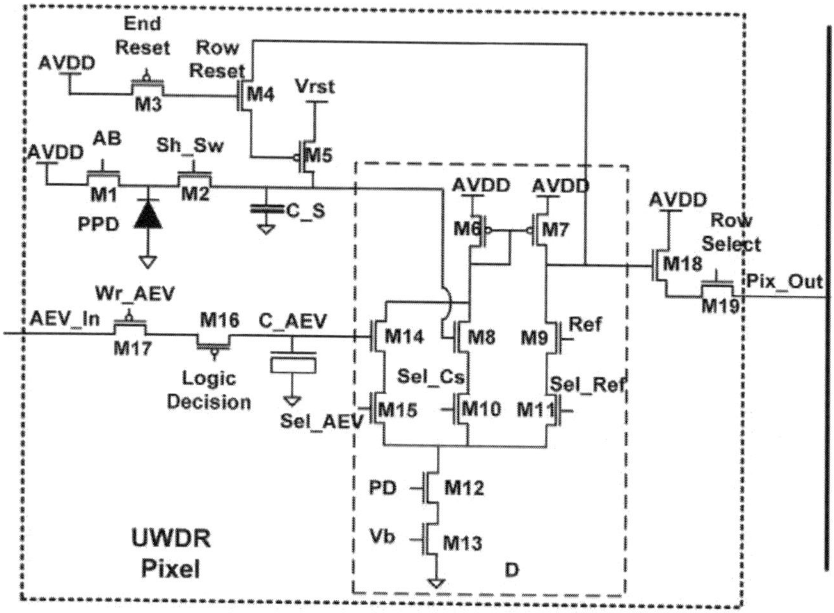

Figure 10. Schematic of **TYPE II** Pixel.

The inputs to this stage are: the pixel integration time represented by *AEV* (M_{14}); the photo-generated signal (M_8); and the global reference signal *Ref* (M_9). These inputs are sampled from C_{AEV}, C_S, and *Ref*, respectively. By means of *Sel_AEV* (M_{15}) or *Sel_*C_s (M_{10}), the differential stage compares the *AEV* or the photo-generated signal, to the *Ref*. the transistor M_{11}, controlled by *Sel_Ref* signal, was added to ensure an optimal matching within the differential amplifier. The bias point is controlled by an analog signal $V_b(M_{13})$, which sets the current magnitude throughout the amplifier branches. To facilitate the power reduction, we inserted an additional transistor M_{12} in the series to the current source in order to shut down the entire stage, when no comparison is needed. The load of the *D* stage is implemented by two PMOS transistors: M_6 and M_7. The drain of the latter is connected to the drain of M_4, forming a self-reset structure.

The self-reset feature of the *D* amplifier enables the application of the conditional reset operation during a saturation check simultaneously to every pixel within the array rather than row by row. At each saturation check, the *Row Reset* (M_4) is raised globally, connecting the output of the *D* stage to the reset transistor M_5. Obviously, when the amplifier's output is low, the C_s capacitance is reset to V_{rst}, otherwiseC_s remains without a change. To stop the reset cycle, regardless of the differential stage output, we use*End Reset* signal (M_3), which forces *AVDD* on the gate of the reset transistor M_5.

AEV assignments are done row by row as will be explained further. Before the assignment operation, the first *AEV* is asserted onto the *AEV_In* bus. Then, the signal *Wr_AEV* (M_{17}) selects the appropriate pixels' row. The *Logic Decision* signal (M_{16}), which is driven by the column-wise common source amplifiers, enables the corresponding *AEV* to be sampled by the C_{AEV} capacitor. The charge flow from the *PPD* is the same as in the **TYPE I** design throughout the whole frame.

At the end of the frame, both the *AEV* and the photo-generated charge are sampled nondestructively through the source follower (SF) (M_{18} and M_{19}) and converted by column-wise amplifiers, as will be explained later.

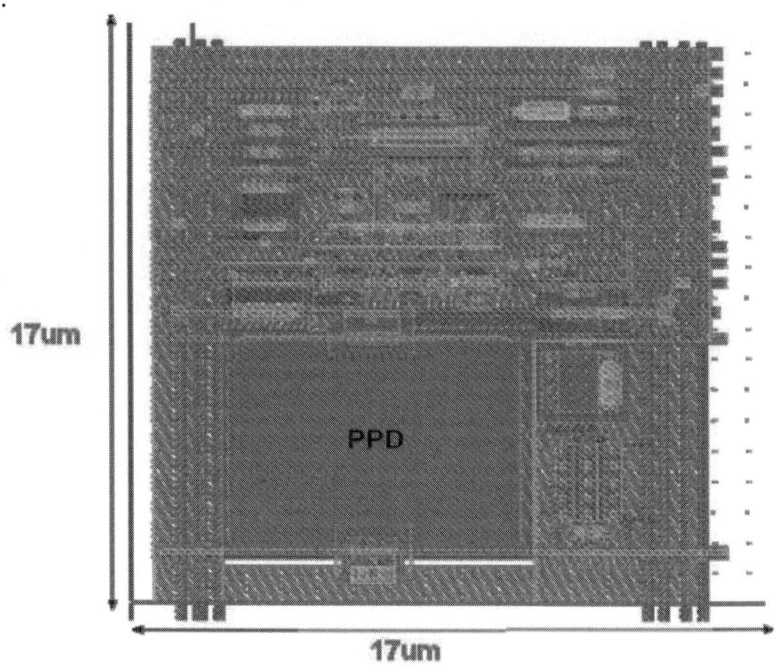

Figure 11. Layout of the **TYPE II** Pixel.

The **TYPE II** pixel was implemented in a 0.18 CMOS process (Figure 11). All the control signals were divided into two groups: 7 analog signals, and 10 logic signals. To reduce the coupling between the control lines, we used extensive orthogonal patterning, so that part of the signals was routed along the X axis, whereas another part was laid along the Y axis. The analog signals were implemented in Metal1 and placed along the Y axis to the left and to the right of the *PPD*. Such an arrangement reduced the coupling between the analog lines themselves and created a substantial separation from the rest logic signals, which were laid along the X axis. Moreover, all the signals, which were running horizontally, were implemented in Metal 3, which reduced the coupling between the pixel analog and logic lines even further. Due to the dense layout, we successfully implemented the described pixel with 17 μm pitch and 25% fill factor FF (Figure 11).

The three phases in the **TYPE II** solution are described in detail below. The *AEV* assignment that occurs in Phase 1 requires a high gain comparator. Since the gain of the in-pixel differential amplifier is not high enough, we added an external common source (*CS*) amplifier (Figure 12). In this way, we obtained the three stage high gain comparator: (1) *D*; (2) *SF*; (3) *CS*, respectively. The *Pix Value* is compared to *Ref* inside each pixel of the selected row. The final amplification is performed by *CS*, which drives the *Logic Decision* signal (Figure 12). This signal stamps the appropriate *AEV* value onto C_{AEV}capacitance inside each pixel. Figure 13 illustrates the flow of the *AEV* assignment to a pixel matrix of 2 × 2 at the first span. Each pixel within the matrix has its own coordinates: the first denotes the row and the second denotes the column. We performed the simulation as if each pixel within this matrix receives a different incoming signal. Therefore, we stimulated the pixels with different current sources, imitating different discharge rates. The *AEV* values were generated by a piecewise ascending ramp voltage ranging from 0.9 V up to 2.6 V, corresponding to 18 different values distanced by 0.1 V from each other. The first span occurs after 100ns. Then, the photo-generated charge is transferred to C_S simultaneously within the whole matrix and is compared row by row with the reference, thus causing the corresponding *AEV* to be written onto C_{AEV} capacitance (Figure 10). Capacitances C_{AEV}<0,0> and C_{AEV}<0,1>, found at the row <0>, are processed first, whereas row <1> capacitances are programmed after the completion of row <0> programming.

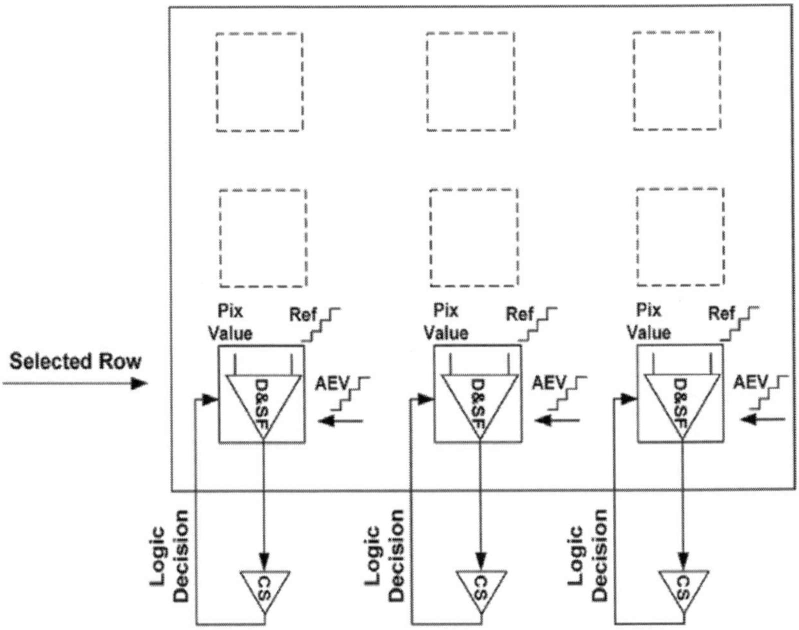

Figure 12. Phase 1 of **TYPE II** Pixel.

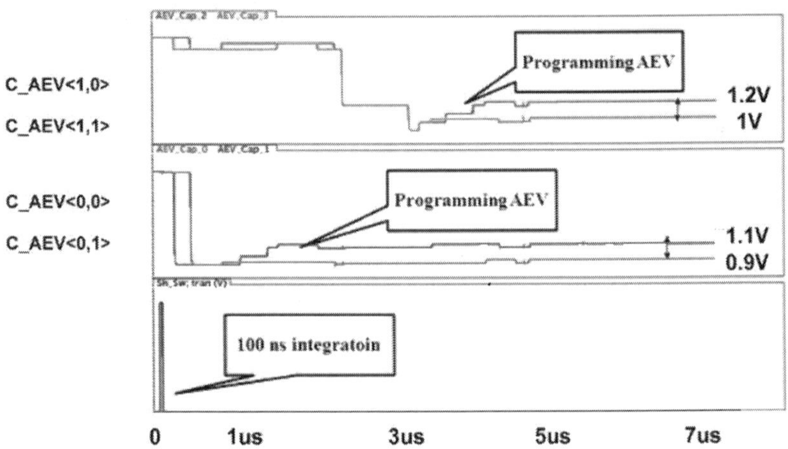

Figure 13. Post-Layout Simulation of Phase 1 of **TYPE II**Pixel.

In the illustrated case, there are four different photo-generated signals, falling within four different segments; therefore every pixel is assigned a different *AEV* from 0.9 V to 1.2 V. The lowest *AEV* is assigned to the most illuminated pixel, *i.e.*, to $C_{AEV}<0,1>$, whereas the rest of the *AEV*'s: 1 V, 1.1 V, and 1.2 V are assigned to less illuminated pixels $C_{AEV}<1,1>$,

C_{AEV}<0,0>, C_{AEV}<1,0>, respectively. It is important to note that the adjacent *AEV* values differ from each other by $2\Delta V_{min}$, which is 0.1 V in this case, and this difference stays intact after the *AEV*'s were written to C_{AEV}. In such a case, decoding the analog data at the end of the frame will be easy; otherwise the final pixel value will be erroneous due to a possible shift of the stored *AEV* values.

It is important to understand that the time of the first span is shorter than the minimal integration time, since, during all the spans, the available signal to be accumulated is lower than the pixel swing by $2\Delta V_{min}$. However, during the next phase the entire pixel swing is available, thus the real integration period will be somewhat higher. In the specific example we have discussed, the minimal integration time will be not 100 ns, but rather 112 ns.

Phase 2 in a **TYPE II** pixel is ultra-fast. Due to its self-reset ability, the reset decision inside each pixel is autonomously fed from the differential stage through M_4 to the gate of M_5 (Figure 10). Consequently, each pixel is reset independently upon the comparison of its own *AEV* with the global reference *Ref*(Figure 14). Therefore, the minimal integration time for this design is:

$$T_{\text{min\_TYPE II}} = T_{AEV\_read} + T_{compare} \quad (6)$$

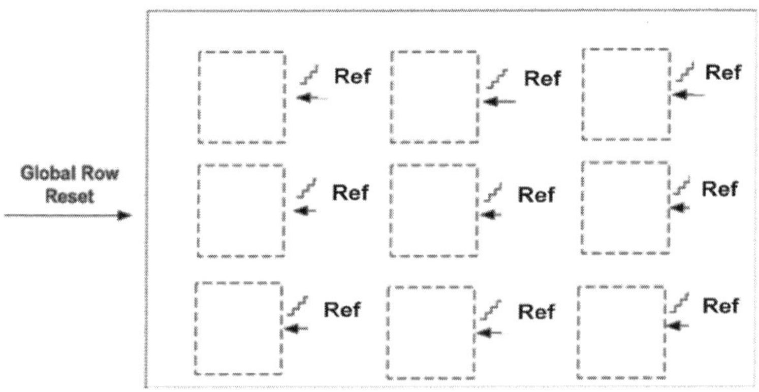

Figure 14. Phase 2 of **TYPE II** Pixel.

Equation (6), which describes the minimal pixel conditional reset time, sets the minimal integration time and, as such, defines the available DR extension. In this case, the minimal integration times can be set as low as 112ns, which enables the DR to be extended by over five decades.

Figure 15a, b illustrates the flow of Phase 2 on a matrix of 2 × 2 pixels. There are four different illuminated pixels, each with its own *AEV* in Phase 1. At Phase 2, every pixel is reset in accordance with its *AEV* (Figure 15a). The bottom pixel, C_S<0, 1>, has the lowest *AEV* and therefore it is reset for five times; whereas the pixel C_S<1,0>with the highest *AEV* is reset only twice. A close–up of the typical reset sequence is presented in Figure 15b. First, the differential stage inside the pixel is activated by raising the*PD* signal; then, the *AEV* is compared to a reference by raising *Sel_AEV*. After this, the bias point of the amplifier stabilizes, the *Row Reset* signal is activated, which enables the decision to reach the reset transistor. In the specific case we have presented, the reset decision is positive and therefore the capacitance *Cs*<0,1> is charged up. The whole sequence lasts for 112 ns; thus, by assuming a typical frame time of 30 ms, we obtain DR extension of 109 dB [8]. Adding to this the intrinsic DR of 60 dBs (10 bits), we obtain overall DR of 170 dB approximately.

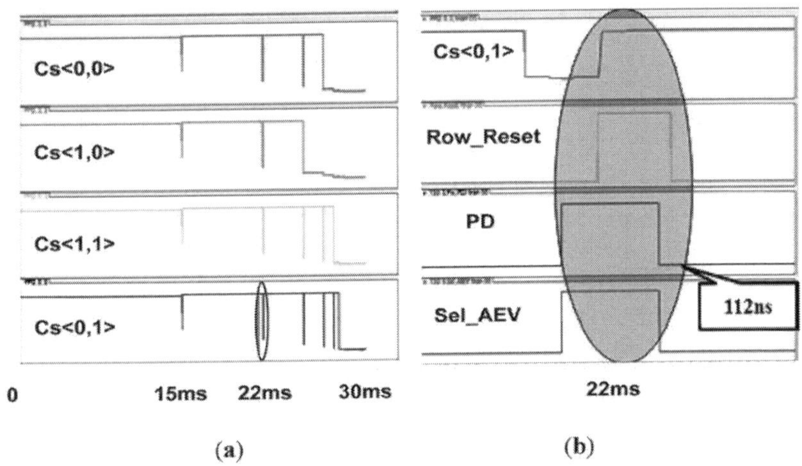

Figure 15. Post-Layout simulation of: (**a**) 4 different pixel signals during Phase 1; (**b**) a close up of a single self-reset cycle.

The final A/D conversion occurs at Phase 3. We use the same comparator configuration as depicted inFigure 12 to perform a single slope conversion to a photo-generated signal (*Mantissa*) and to an

AEV.Figure 16 depicts the process of A/D conversion to the stored *AEV* and the pixel *Mantissa*. First, the pixel*AEV* is converted using a ramp with reduced ("coarse") resolution. It can easily be observed that the *Ref*signal spans the *AEV* domain in a reduced number of steps, which corresponds to the "coarse" conversion; whereas, the pixel *Mantissa* is spanned by the reference with a much higher ("fine") resolution. As a matter of fact, after the *Mantissa* is converted, the capacitance C_S storing it is being reset for the next frame. The capacitance holding the *AEV* is not necessary to be reset, since it will be assigned the new *AEV* in Phase 1 of the subsequent frame in any case.

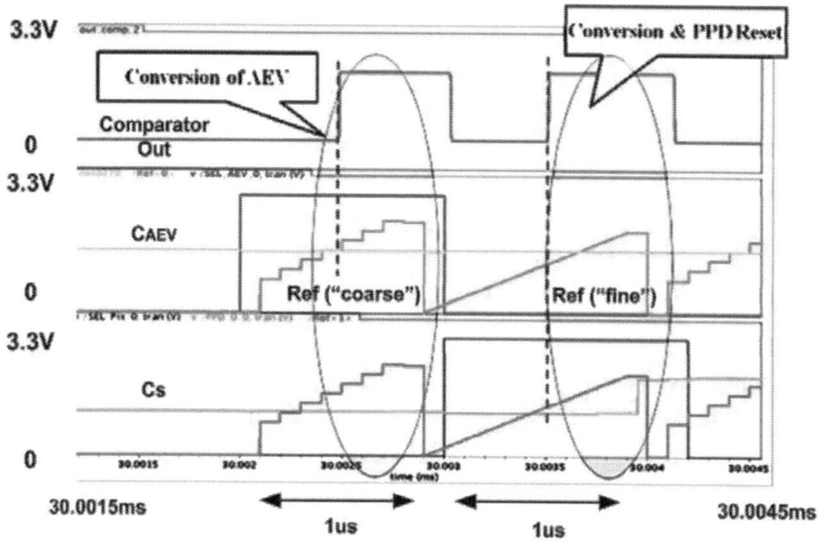

Figure 16. Post-Layout Simulation of Phase 3 of **TYPE II**pixel.

In conclusion, the relatively complex structure of the **TYPE II** pixel has enabled a very large DR extension and has substantially simplified the second and the third phases of the proposed algorithm.Figure 15, Figure 16 prove the feasibility of the **TYPE II** design and indicate that sensor based upon a**TYPE II** pixel can be successfully implemented in a silicon process.

A Power Profile of TYPE I and TYPE II Designs

Another important factor is the examination of the power performance of the **TYPE I** and **TYPE II**designs, since, if they were power hungry, it would be pointless using them. We made both of the designs optimized for low power using the following methods: the two stacked transistors scheme, and multiple power supplies. The result was that the leakage within the pixels was successfully minimized, and the power supplies could be lowered easily without degrading the anticipated sensor's performance.

In analyzing the **TYPE I** power per pixel (*PPX*), we have concluded that most of the power is consumed during the *AEV* assignment (Phase 1) and the A/D conversion (Phase 3), because these two operations require activating the comparators for a longer period of time than in Phase 2 (Figure 17a).

In **TYPE II**, the relative power contribution of Phase 2 is also the lowest, in comparison with the two other phases, as its overall effective duration is the shortest one (Figure 17b). Most of the power is consumed during Phase 1, which contains an increased number of spans and, thereby, in aggregate, lasts longer than Phase 3. Phase 3's contribution to *PPX*, is almost the same as in the **TYPE I** case due to similar flow of the data conversion.

Based on these simulations, we have concluded that the two presented designs consume power of 2.5 nW and 5.4 nW, respectively. Taking into account the anticipated extraordinary DR extension they provide, we have found that the power budget is low and definitely appropriate to the state-of-the art CMOS image sensors [25].

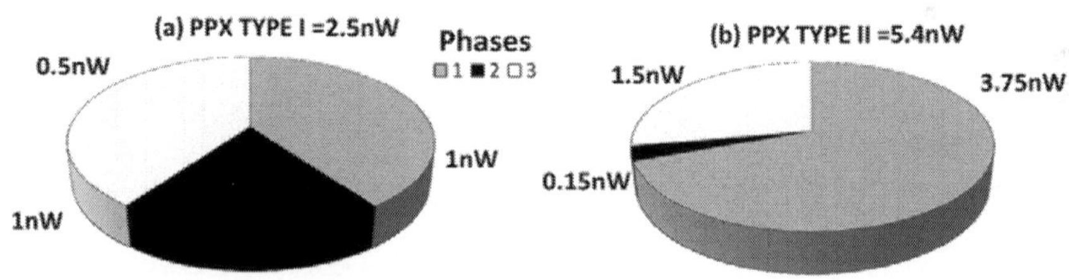

Figure 17. Power per Pixel Distribution of **TYPE I** and **TYPE II** Designs. (**a**) PPX TYPE I = 2.5 nW; (**b**) PPX TYPE I = 5.4 nW.

To summarize the anticipated performance of the proposed designs, we provide Table 1, where different state-of-the art WDR solutions are compared with respect to several key attributes such as fabrication technology, WDR technique (WDR T.), pixel size, fill factor (FF), DR, SNR, power per pixel (PPX), and frame rate (FR).

Table 1. Comparison between state-of-the art WDR image sensors.

Parameter	[11]	[5]	[12]	[14]	[21]	TYPE I	TYPE II
Technology	0.18 µm	0.09 µm	0.18 µm	0.35 µm	0.18 µm	0.18 µm	0.18 µm
WDR T.	Well Cap.Adjustment	Well Cap.Adjustment	Frequency	TFS	Multiple Capt.	Multiple Resets	Multiple Resets
Pixel Size	3 µm × 3 µm	5.86 µm × 5.86 µm	23 µm × 23 µm	81.5 µm × 76.5 µm	5.6 µm × 5.6 µm	14 µm × 14 µm	17 µm × 17 µm
FF	-	-	25%	2%	45%	40%	25%
DR	100 dB	83 dB	130 dB	100 dB	99d B	108 dB	170 dB
SNR	48 dB*	48 dB *	-	-	-	48 dB *	48 dB *
PPX	-	400 nW **	250 nW	≤6.4 µW **	10 nW **	2.5 nW	5.4 nW
FR	-	30	-	-	15	33	33

From Table 1, we can understand that the pixel sizes of pixels presented in our work are mediocre. They are substantially larger than those associated with well capacity adjustment and multiple captures, but much smaller than frequency and TFS based sensors. Important to note that both of our designs maintain a decent FF relatively other listed solutions due to area effective layout. The DR of proposed herein designs emphasizes the extraordinary ability of multiple resets algorithm to extend the pixel dynamic range. **TYPE I** provides remarkable DR, whilst **TYPE II** brings it to extreme values. Moreover, such DR is obtained, maintaining excellent SNR (see Appendix B). Not less important to note that both of proposed designs operate at video frame rate and present the best power performance.

A COMPARATIVE ANALYSIS OF MULTIPLE CAPTURES & MULTIPLE RESETS ALGORITHMS

When a new WDR solution is proposed, it is always interesting to compare it to the existing ones and to realize the added value the new solution contributes. In the present discussion, we perform a comparison between the two proposed solutions to the following algorithms: the multiple captures [15] and the former multiple resets solutions [24]. We chose to compare the number of operations needed to obtain a single digitized WDR pixel value; because, in our opinion, the most significant contribution of**TYPE I** and **TYPE II** solutions is their effectiveness and their ability to substantially reduce the processing time.

First, the multiple captures algorithm must be considered. In this solution, the pixel integrates for several periods of time, regardless of the intensity of the incoming light. After each integration, the intermediate pixel value is synthesized out of both the newly generated and previously stored samples, respectively. At this stage we assume that all the captures are performed in a rolling shutter mode to comply with the minimal memory requirements for this solution. In case multiple captures are performed in a global shutter mode, the memory requirements will be much higher. For a rolling shutter operation mode, the number of memory bits for each pixel equals its digital resolution: $N_{resolution}$. The total number of operations that are required to obtain a final digitized result for a single pixel after $N_{captures}$ exposures is:

$$N_{op\_captures\_tot} = \left(N_{captures} - 1\right)\left(1_{convert} + N_{resolution}\left(1_{read} + 1_{write}\right)\right) \tag{7}$$

where $1_{convert}$, 1_{read}, and 1_{write} stand for ADC, memory read, and write cycles, respectively. Please note, that in this case there are two captures only: the memory actually becomes a single row register where the first sample is kept, in order to be compared with the second.

On the other hand, both the previous and newly proposed multiple resets solutions are applied on a global shutter sensor to obtain the maximal possible number of operations. According to the snapshot mode, the number of memory bits for a single pixel equals Rep_{EXP}, which is the representation of the exponent bits N_{EXP}. Obviously, in case of linear representation, Rep_{EXP} equals N_{EXP}. If logarithmic representation is applied, the Rep_{EXP} is reduced to $\log_2 N_{EXP} + 1$ [27]. For example, if there are 6 exponent bits, then in the first case, Rep_{EXP} is 6, whereas in the second it is 4. The overall number of operations in previous solutions is:

$$N_{op\_resets\_tot\_prev} = N_{EXP}\left(1_{read} + 1_{convert} + 1_{write}\right) + 1_{read} Rep_{EXP} + 1_{convert} \tag{8}$$

The first term in (8), relates to the sequential generation of DR bits, which consists of memory read, pixel conversion by the comparator, and memory write cycles. The second term relates to the digital readout of the memory contents, and the last term denotes the A/D conversion of the pixel *Mantissa* at the end of the frame.

In the newly presented **TYPE I** and **TYPE II** solutions, the total number of operations coincides with:

$$N_{op\_resets\_tot\_newv} = 1_{convert}\left(l + N_{EXP} + 2\right) \quad (9)$$

The term l stands for the number of spans performed during Phase 1; N_{EXP} is for Phase 2, and the last term of 2 denotes the number of conversions during Phase 3.

Next, we will demonstrate the meaning of Equations (7–9) by substituting the possible values for each parameter. We have presented the results in Table 2, where each row is allocated for a different $N_{captures}$and N_{EXP}.

For the previous solutions of multiple resets, we chose the logarithmical representation as in [27], because it minimizes the required external memory size. We also have assumed that the number of the exponent bits N_{EXP} equals the number of captures: $N_{captures}$. This assumption not only simplifies the calculations, but it also ensures that the SNR dips, which exist throughout the extended DR in all of the solutions, are equal.

The advantage of the multiple captures algorithm is that most of it is implemented in a digital domain and the pixel structure is rather simple. The clear disadvantage is that this solution requires an extensive processing to obtain the final result. From Equation (7), it is shown that the number of operations increases linearly with the number of captures. This fact is clearly demonstrated by Table 2, where the multiple captures solution obviously possesses the highest number of operations per pixel. It is possible, though, to limit the number of captures to two or three; however this will cause large SNR dips at the boundary light intensities [8]. Circumventing these SNR artifacts by adding more integration slots will certainly slow the sensor operation speed. Consequently, the multiple captures algorithm is best suited to photograph slow changing scenes with a very high spatial resolution.

Table 2. The Comparison of a number of operations for "Multiple Captures", "Multiple Resets Previous", and "Multiple Resets New" algorithms.

	$N_{EXP} = N_{captures}$		
$N_{captures} = N_{EXP}$	**Multiple Captures [15]**	**Multiple Resets Previous [27]**	**Multiple Resets New (This Work)**
2	21	9	5
3	42	13	6
4	63	16	7
6	105	23	10
18	357	61	25

In the multiple resets algorithm, although the pixel structure and the pixel control are more complex than in the first case, the overall processing is much faster, since the pixel adjusts the integration time autonomously during the frame. Thus, there is no need for an extensive post-integration processing—only to combine the data of the *Mantissa* and the *EXP* bits. From Table 2, we realize that even the previous configurations of multiple resets have a reduced number of operations, relative to the multiple captures algorithm. The intuitive explanation of this is found within a sequential processing of the pixel information, which allows reading only the last produced bit from the memory before generating the subsequent one. The logarithmical representation of the *EXP* bits, used in the previous designs, decreases the operations number even further. Therefore, the previous solutions of multiple resets were perfectly suited to capture fast changing WDR scenes with a decent resolution.

The two solutions, proposed within the current study, bring the number of operations to its minimum, since they completely eliminate the operations related to a digital memory unit. This feature enables the proposed solutions to be the fastest among the three, which have been analyzed in this discussion. The clear disadvantage of the pixel using *AEV* is a relatively complex analog signal processing inside each pixel, which is especially pronounced in **TYPE II** design. Therefore, the newly proposed solutions can be candidates for implementation for capturing a very fast changing scene with wide or ultra-wide dynamic range.

SUMMARY

We have presented two designs: **TYPE I** and **TYPE II** featuring the "ranging" and the *AEV* concepts. We have shown that it is possible to perform effective encoding of the DR bits using analog voltage, which can be easily stored within the pixel and processed further through the readout chain. Moreover, we have condensed the *AEV* assignment to a single phase in a process called "ranging", which has led to another improvement: a substantial simplification of the conditional reset. In the past, each cycle consisted of three or even more different operations; now the number of operations has been nearly halved. The *AEV* solution has also granted a possibility to save chip area due to reuse of high gain comparators as a main block in a single slope ADC. The post-layout simulations and power calculations included in our study have shown that both the **TYPE I** and the **TYPE II** designs are feasible and can be successfully implemented in a CMOS process. The presented designs have reached a remarkable DR of up to 170 dB, while only consuming power of up to 5.4 nW. A further comparison between this work and the multiple captures and the former multiple resets solutions has clearly proven that the *AEV* concept effectively reduces the number of operations required to produce a pixel value and, as such, can be considered as a key to design a fast, low-power and ultra WDR CMOS sensor.

APPENDIX A

Division of the Pixel Swing to Segments

Since our algorithm is based upon integration times ordered in geometrical progression, we divide the pixel swing S_{pix} into geometrically ordered segments as well: S_1, S_2, ..., S_n as depicted in Figure 18.

During the frame, the pixel discharges from the *Reset level* towards the noise floor. The *Reset level* and the lower bound of pixel signal define the pixel swing S_{pix}. The first segment S_1 equals $2\Delta V_{min}$. Consequently, the subsequent segments will be equal to $4\Delta V_{min}$, $8\Delta V_{min}$, and so on, till the last segmentS_n, (Figure 18). In the general form, the values of these segments are given by:

$$S_i = 2^{i-1} \cdot (2\Delta V_{min}) \leq S_{pix} - 2\Delta V_{min} \quad 1 \leq i \leq n \tag{10}$$

It is important to note that the last segment S_n, is distanced from the pixel swing S_{pix} by $2\Delta V_{\min}$. This means that signals exceeding S_n will not be converted. The purpose of this restriction will be explained later on.

That the number of *AEV* values that can be assigned in a single span equals the number of segments, which are given from (10) can be easily understood. For example, as shown in (Figure 18) if the pixel $S_n$equals $16\Delta V_{\min}$, there will be four different segments; thus, four values of *AEV*s can be assigned during a single conversion.

The ramp signal *Ref* spans the pixel signal. This signal has to comply with the following requirements: (1) the signals that reach the lower bound of S_{pix} will not be converted; and (2) signals reaching the edge of the last segment, S_4 in this case, will be converted. In this way, the ramp starts on its way in the middle of the restricted region marked in red (Figure 19).

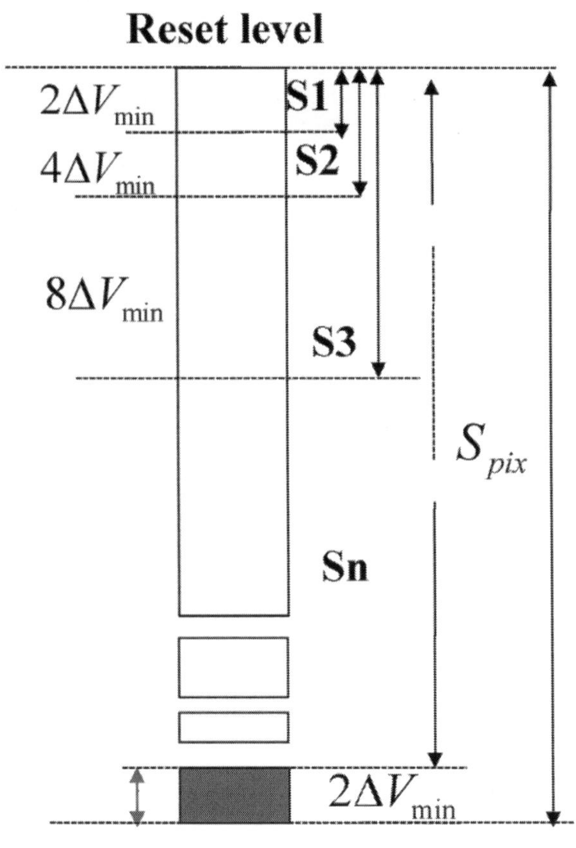

Figure 18. Division of the Pixel Swing to Segments.

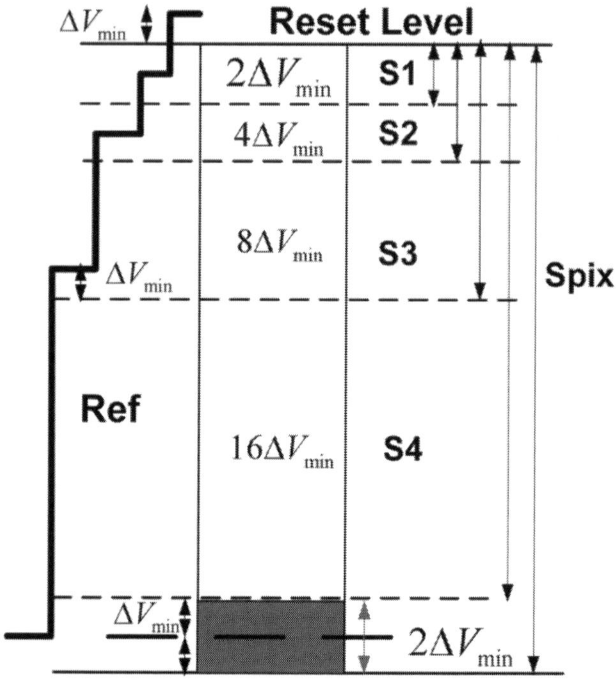

Figure 19. Conversion of a pixel swing composed out of four segments.

From there, it climbs in geometrically decreasing steps, gradually covering all the pixel segments. To insure that all the segments are duly covered, the ramp *Ref* surpasses each one by ΔV_{min}. An additional positive consequence of such ramp configuration is that the minimal SNR during each conversion is set by ΔV_{min} (see Appendix B).

Special attention needs to be paid to the synchronization between the change of *AEV* and the ramp steps (Figure 20). Please note that *AEV* 2 becomes valid before the *Ref* continues to climb and it stays constant for the $t_{program}$ interval, during which the *AEV* 2 value is being written onto the C_{AEV}.

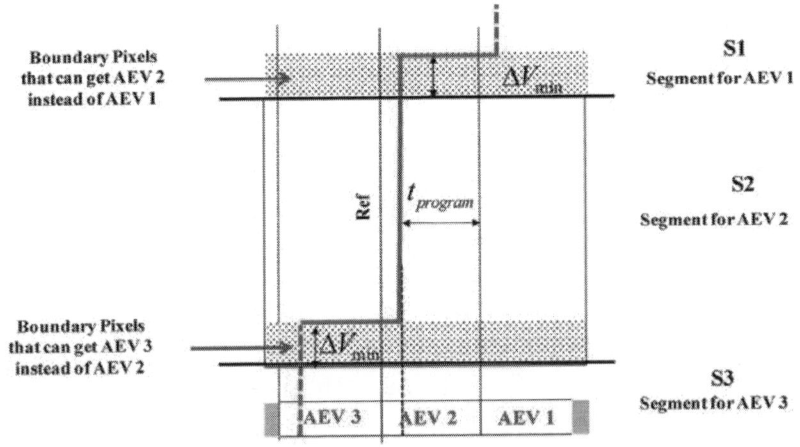

Figure 20. *AEV* change in correlation with the comparator reference.

Another essential aspect is the assignment of the *AEV* values to the pixels, the signals of which lie at the boundaries of the adjacent segments. Such *Boundary Pixels* can receive the *AEV* value corresponding to the previous segment. For example, the ramp *Ref* finalizes spanning the pixel signals that should receive *AEV* 2 within the adjacent S_1 segment. In such a situation, some pixels, found at the lower boundary of a S_1 segment can mistakenly receive *AEV* 2 instead of *AEV* 1. As a result, the pixels' saturation is prevented, but this is at the expense of the reduction of half of their effective integration time, which leads to a SNR drop of 6 dBs.

Up to this point, we have explained the aspects of a single pixel span, but what happens if we were to perform additional spans to complete the desired number of *EXP* bits N_{EXP}? Then, the number of spans l is increased to (11). For example, if we suppose there are 18 bits to be encoded into 18 *AEV* values and in each span 4 bits are encoded, then 5 scans are needed to complete Phase 1.

$$l = \left\lceil \frac{N_{EXP}}{n} \right\rceil \tag{11}$$

where n is the number of segments. Since there might be several spans, how are we to ensure that *AEV*s, which were assigned in the preceding spans, are not overridden in the subsequent ones? The answer is to distance the last segment by $2\Delta V_{min}$ from the pixel lower boundary. If we look for the reason the *AEV*'s were not overridden, we learn that there are two spans necessary to encode eight different light intensities "1–8" (Figure 21a,b). After the first span, the four most significant intensities "5–8" are covered and assigned their *AEV*'s (Figure 21a). The weaker four intensities ("1–4") are assigned *AEV*'s, which are not correct and therefore should be overwritten in the upcoming second span. During the second span, all the pixels will be addressed once again, regardless if they have already received their *AEV*'s or not. The second span is distanced from the first, so that the signal generated by "5–8" exceeds the last segment by $2\Delta V_{min}$. Therefore, these pixels do not receive new *AEV*'s. On the other hand, "1–4" intensities generate signals, which now fall within the different segments and thereby are assigned the corresponding *AEV*'s (Figure 21b). Hence, by the reduction of the possible conversion range, we can differentiate between the already encoded pixels and those which should be encoded in the future spans.

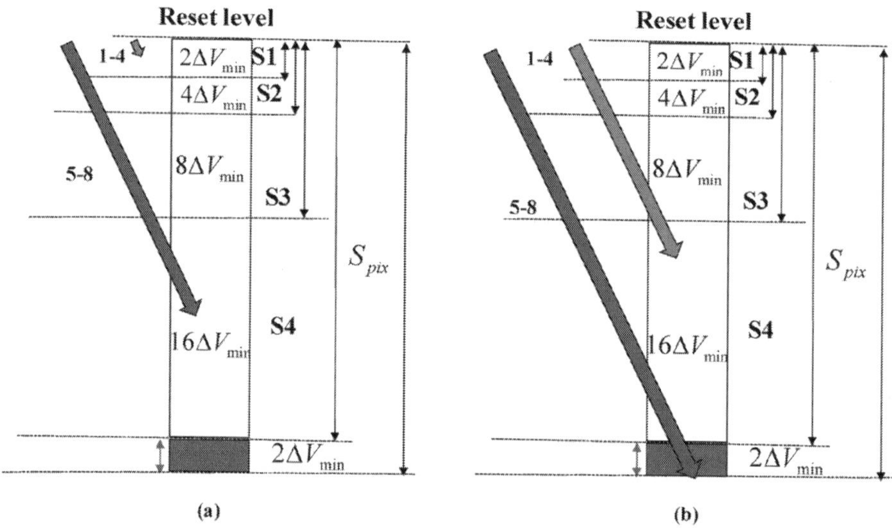

Figure 21. Distribution of the different light intensities by segments (**a**) first span; (**b**) second span.

All the spans are spread non-uniformly in time. The occurrence of each span t_k is set upon the most significant *EXP* bit that can be encoded in the *k*-th span:

$$t_k = \frac{T_{int}}{2^{M_k}} \cdot \frac{S_{pix} - 2\Delta V_{min}}{S_{pix}} \quad k = 1, 2, \ldots, l \tag{12}$$

where M_k is the MSB that can be encoded in the k-th span. For example, if there are 18 *EXP* bits to be encoded during 5 spans, then $M_1 = 18$; $M_2 = 14$; $M_3 = 10$; $M_4 = 6$; $M_5 = 2$, respectively. Furthermore, assuming that $S_{pix} = 1$ V; $2\Delta V_{\min} = 0.112$ V; $T_{int} = 30$ ms, the span times t_1 till t_5 are: 100 ns; 1.65 μs, 26.4 μs, 422 μs, and 6.75 ms, respectively

APPENDIX B

SNR Considerations

SNR is one of the key properties of every sensor. Every designer must be aware how much noise each specific circuitry architecture generates. Herein, we discuss three noise sources shot noise, flicker noise, and the thermal noise. We consider the noise of the two types of the sensors separately. The performed analysis shows that in poorly illuminated scenes, the thermal noise governs the noise floor for both of the sensors, whereas the shot noise of photo-diode sets the noise limit in highly illuminated scenes. We refer to noise induced to *AEV* capacitance as well, and show that our design considerations provide an excellent signal to noise ratio of the stored data throughout the whole frame.

TYPE I

We assume, for our convenience that all noise sources are uncorrelated, therefore we can write the overall SNR as:

$$SNR = 20\log\frac{i_{ph}t_{\text{int}}}{\sqrt{V^2_{n\_KTC} + V^2_{n\_shot} + V^2_{n\_read}}} \tag{13}$$

where V_{n_KTC}, V_{n_shot}, V_{n_read} are the voltage variances of the thermally generated KTC noise, shot noise, and read noise, respectively. i_{ph} denotes the photo-generated current and t_{int} denotes the time it is integrated within the photo-diode.

The noise analysis for this sensor is straightforward, due to its conventional structure. At the beginning of the frame, the in-pixel sense capacitance C_S (Figure 5) is being reset, which causes well-known KTC noise. Since it is "hard reset" [28], performed through PMOS, the overall inflicted noise equals:

$$V_{n\_KTC} = \sqrt{\frac{k_B T}{C_S}} \tag{14}$$

where k_B and T are the Boltzman's constant and junction temperature in degrees of Kelvin, respectively. According to our assessments, the resulted noise equals to 643 μV. Note, that since at the end of the frame, we do not need to convert the reset signal as in other conventional designs, this noise is not doubled. In case, there are multiple resets applied, this noise contribution grows by the square root of the reset times.

Shot noise within the photodiode is inflicted by both the dark current i_{dc} and the incoming optical signal i_{ph} [28]. The overall noise contribution is given by:

$$V_{n\_shot} = \sqrt{\frac{q(i_{ph} + i_{dc})t_{\text{int}}}{C_{PD}^2}} \tag{15}$$

In case the accumulated charge is maximal, *i.e.*, it equals the pixel swing S_{pix}, we get the following noise:

$$V_{n\_shot\_max} = \sqrt{\frac{qS_{pix}}{C_{PD}}}, \quad SNR_{max} = 20\log\frac{S_{pix}}{V_{n\_shot\_max}} = 20\log\sqrt{\frac{C_{PD}S_{pix}}{q}} \tag{16}$$

According to our assessments this amount of noise can reach 4 mV. In such a case the maximal SNR according to (16) equals 48 dB.

The readout noise in this sensor is mainly generated by the in-pixel source follower amplifiers, through which the readouts of *AEV* and the pixel are performed. It is easy to show that the noise source follower amplifier generates [29] is given by:

$$V_{n\_readout\_SF} = \sqrt{\frac{2k_BT}{3C_{Pix\_Out}}} \quad or \quad \sqrt{\frac{2k_BT}{3C_{AEV\_Out}}} \tag{17}$$

where C_{Pix_Out}, C_{AEV_Out}, are the sample capacitances for pixel and the *AEV* readout, respectively. One should note that since the gain of source follower never exceeds unity, the load capacitances should be large enough to reduce the readout noise to desirable level. In our case, the sample capacitances were designed to be no less than 100 fF, which bounded the readout noise to 200 μV, approximately.

It must be noted that there are additional noise sources within the source follower amplifier such as: flicker and telegraph noise [28]. The power of these noise sources is strongly depends on the process quality. With the recent fabrication technology advances, the total contribution of theses noise sources has been reduced to couple of electrons rms, which in our case equal to couple of tens micro-volts.

Additional noise sources, which have to be taken into consideration during the readout, is the thermal noise generated by the column-wise high gain comparator (Figure 7). The comparator consists of two stages: differential and common source. It is easy to show that the input referred thermal noise voltage variance is given by [29]:

$$V_{n\_readout\_Diff} = \sqrt{\frac{4k_BT}{3C_{CS\_in}G_{Diff}}} \tag{18}$$

where C_{CS_in} is the common source input capacitance and G_{Diff} is the differential stage gain. Correspondingly, the noise generated by the common source is:

$$V_{n\_readout\_Diff} = \sqrt{\frac{2k_BT}{3C_{CS\_Out}G_{Diff}G_{CS}}} \tag{19}$$

Substituting the corresponding values into (18), (19), we assessed that the input referred noise coming from the comparator is 55 μV, which is substantially lower than the one induced by the source follower.

The in-pixel dynamic capacitance storing *AEV* value also accommodates a certain amount of noise. Apparently, it is the KTC noise, which is injected onto C_{AEV} (Figure 5) the moment the *Logic Decision*signal disconnects it from the *AEV_In* bus. The amount of noise can be calculated by (14) by substitutingC_{AEV} value into that formula. Since C_{AEV} is almost the same as C_S the noise equals 643 μV as well. Taking into account that the minimal difference between the *AEV* value and the reference signal *Ref*(Figure 19) is $\Delta V_{\min}$, we get the following minimal SNR:

$$SNR_{min} = 20\log_{10}\sqrt{\frac{C_{AEV}\Delta V_{min}}{k_BT}} \tag{20}$$

For example, if C_{AEV} is 10 fF and ΔV_{min} is 50 mV, we can observe that the minimal SNR is 37 dB, which is enough for obtaining 4–5 bits resolution during each span.

To conclude, **TYPE I** sensor reaches 48 dB SNR, when exposed to high light intensities. The dominant noise at low light is the KTC, whereas at high end it is the shot noise, which limits the sensor performance.
TYPE II

The SNR in this sensor is also given by (13). The first two terms of noise, namely V_{n_KTC}, V_{n_shot}, are similar to the **TYPE I** sensor. The readout noise, however, is much lower due to accommodation of the differential stage inside the pixel. In this configuration, the noise contributions of source follower, common source become reduced by the gain of the in-pixel differential stage. Thus, employing Equations (17)–(19), we can assess that the input referred readout noise does not exceed 55–60 μV. Obviously, the noise floor of **TYPE II** varies almost the same as of **TYPE I** throughout the whole illumination range.

The noise associated with *AEV* is governed by the KTC noise injected during Phase 1 and therefore equals the one for **TYPE I** sensor.

REFERENCES

1. Mizobuchi, K.; Adachi, S.; Tejada, J.; Oshikubo, H.; Akahane, N.; Sugawa, S. A low-noise wide dynamic range CMOS image sensor with low and high temperatures resistance. *Proc. SPIE* **2008**,*6816*, 681604:1–681604:8.
2. Stevens, E.; Rochester, N.Y. Low-Crosstalk and Low-Dark-Current CMOS Image-Sensor Technology Using a Hole-Based Detector. In Proceedings of the ISSCC—Image Sensors and Technology, San-Francisco, CA, USA, 3–7 February 2008; pp. 60–62.
3. Shcherback, I.; Belenky, A.; Yadid-Pecht, O. Empirical dark current modeling for complementary metal oxide semiconductor active pixel sensor. *Opt. Eng.* **2002**, *41*, 1216–1219.
4. Kavadias, S.; Dierickx, B.; Scheffer, D.; Alaerts, A.; Uwaerts, D.; Bogaerts, J. A logarithmic response CMOS image sensor with on-chip calibration. *IEEE J. Solid-State Circuits* **2000**, *35*, 1146–1152.
5. Sakakibara, M.; Oike, Y.; Takatsuka, T.; Kato, A.; Honda, K.; Taura, T.; Machida, T.; Okuno, J.; Ando, A.; Fukuro, T.; *et al.* An 83dB-Dynamic-Range Single-Exposure Global-Shutter CMOS Image Sensor with In-Pixel Dual Storage. In Proceedings of the IEEE International Solid State Circuits Conference Digest Technical Papers (ISSCC), San Francisco, CA, USA, 19–23 February 2012; pp. 380–382.

6. Seo, M.W.; Suh, S.; Iida, T.; Watanabe, H.; Takasawa, T.; Akahori, T.; Isobe, K.; Watanabe, T.; Itoh, S.; Kawahito, S. An 80 μV_{rms}-Temporal-Noise 82 dB-Dynamic-Range CMOS Image Sensor with a 13-to-19b Variable-Resolution Column-Parallel Folding-Integration/Cyclic ADC. In Proceedings of the IEEE International Solid-State Circuits Conference Digest of Technical Papers (ISSCC), San Francisco, CA, USA, 20–24 February 2011; pp. 400–402.
7. Lu, J.H.; Inerowicz, M.; Joo, S.; Kwon, J.K.; Jung, B. A low-power, wide-dynamic-range semi-digital universal sensor readout circuit using pulse-width modulation. *Sens. J. IEEE* **2011**, *11*, 1134–1144.
8. Spivak, A.; Belenky, A.; Fish, A.; Yadid-Pecht, O. Wide-dynamic-range CMOS image sensors—Comparative performance analysis. *Electron. Devices IEEE Trans.* **2009**, *56*, 2446–2461.
9. Storm, G.; Henderson, R.; Hurwitz, J.E.D.; Renshaw, D.; Findlater, K.; Purcell, M. Extended dynamic range from a combined linear–logarithmic CMOS image sensor. *IEEE J. Solid State Circuits* **2006**, *41*, 2095–2106.
10. Akahane, N.; Sugawa, S.; Adachi, S.; Mori, K.; Ishiuchi, T.; Mizobuchi, K. A sensitivity and linearity improvement of a 100-dB dynamic range CMOS image sensor using a lateral overflow integration capacitor. *IEEE J. Solid State Circuits* **2006**, *41*, 851–858.
11. Sakai, S.; Tashiro, Y.; Kawada, S.; Kuroda, R.; Akahane, N.; Mizobuchi, K.; Sugawa, S. Pixel scaling in complementary metal oxide silicon image sensor with lateral overflow integration capacitor. *Jpn. J. Appl. Phys.* **2010**, *49*, 1–6.
12. Wang, X.; Wong, W.; Hornsey, R. A high dynamic range CMOS image sensor with in-pixel light-to-frequency conversion. *IEEE Trans. Electron. Devices* **2006**, *53*, 2988–2992.
13. Kitchen, A.; Bermak, A.; Bouzerdoum, A. PWM digital pixel sensor based on asynchronous self-resetting scheme. *IEEE Electron. Device Lett.* **2004**, *25*, 471–473.
14. Leñero-Bardallo, J.A.; Serrano-Gotarredona, T.; Linares-Barranco, B. A five-decade dynamic-range ambient-light-independent calibrated signed-spatial-contrast AER retina with 0.1-ms latency and optional time-to-first-spike mode. *Circuits Syst. I IEEE Trans.* **2010**, *57*, 2632–2643.
15. Mase, M.; Kawahito, S.; Sasaki, M.; Wakamori, Y.; Furuta, M. A wide dynamic range CMOS image sensor with multiple exposure-time signal outputs and 12-bit column-parallel cyclic A/D converters. *Solid State Circuits IEEE J.* **2005**, *40*, 2787–2795.
16. Dattner, Y.; Yadid-Pecht, O. High and low light CMOS imager employing wide dynamic range expansion and low noise readout. *Sens. J. IEEE* **2012**, *12*, 2172–2179.
17. Sandhu, T.S.; Pecht, O.Y. New memory architecture for rolling shutter wide dynamic range CMOS imagers. *Sens. J. IEEE* **2012**, *12*, 767–772.
18. Woo, D.H.; Nam, I.K.; Lee, H.C. Smart reset control for wide-dynamic-range LWIR FPAs. *Sens. J. IEEE* **2011**, *11*, 131–136.
19. Stoppa, D.; Vatteroni, M.; Covi, D.; Baschirotto, A.; Sartori, A.; Simoni, A. A 120-dB dynamic range CMOS image sensor with programmable power responsivity. *IEEE J. Solid State Circuits***2007**, *42*, 1555–1563.

20. Yasutomi, K.; Itoh, S.; Kawahito, S. A two-stage charge transfer active pixel CMOS image sensor with low-noise global shuttering and a dual-shuttering mode. *Electron. Devices IEEE Trans.* **2011**,*58*, 740–747.
21. Choi, J.; Park, S.; Cho, J.; Yoon, E. A 1.36μW adaptive CMOS Image Sensor with Reconfigurable Modes of Operation from Available Energy/Illumination for Distributed Wireless Sensor Network. In Proceedings of the 2012 IEEE International Solid-State Circuits Conference Digest of Technical Papers (ISSCC), San Francisco, CA, USA, 19–23 February 2012; pp. 112–114.
22. Han, S.W.; Kim, S.J.; Choi, J.H.; Kim, C.K.; Yoon, E. A High Dynamic Range CMOS Image Sensor with In-Pixel Floating-Node Analog Memory for Pixel Level Integration Time Control. In Proceedings of the Symposium on VLSI Circuits Digest of Technical Papers, Hilton Hawaiian Village, Honolulu, HI, USA, 13–15 June 2006; pp. 25–26.
23. Snoeij, M.F.; Theuwissen, A.J.P.; Makinwa, K.A.A.; Huijsing, J.H. Multiple-ramp column-parallel ADC architectures for CMOS image sensors. *Solid State Circuits IEEE J.* **2007**, *42*, 2968–2977.
24. Spivak, A.; Teman, A.; Belenky, A.; Yadid-Pecht, O.; Fish, A. Power-performance tradeoffs in wide dynamic range image sensors with multiple reset approach. *J. Low Power Electron. Appl.***2011**, *1*, 59–76.
25. Spivak, A.; Teman, A.; Belenky, A.; Yadid-Pecht, O.; Fish, A. Low-voltage 96 dB snapshot CMOS image sensor with 4.5 nW power dissipation per pixel. *J. Sens.* **2012**, *12*, 10067–10085.
26. Yadid-Pecht, O.; Fossum, E.R. Image sensor with ultra-high-linear dynamic range utilizing dual output CMOS active pixel sensors. *IEEE Trans. Electron. Devices* **1997**, *44*, 1721–1724.
27. Belenky, A.; Fish, A.; Spivak, A.; Yadid-Pecht, O. Global shutter CMOS image sensor with wide dynamic range. *IEEE Trans. Circuits Syst. II Exp. Briefs* **2007**, *54*, 1032–1036.
28. Tian, H.; Fowler, B.; Gamal, A.E. Analysis of temporal noise in CMOS photodiode active pixel sensor. *Solid State Circuits IEEE J.* **2001**, *36*, 92–101.
29. Razavi, B. *Design of Analog CMOS Integrated Circuits*; McGraw-Hill Higher Education: New York, NY, USA, 2001; pp. 209–240.
30. Janesick, J.R. *Photon Transfer DN to λ*; SPIE Library: Bellingham, WA, USA, 2007; pp. 163–164

CITATION

Arthur Spivak , Alexander Belenky , Alexander Fish and Orly Yadid-Pecht, Analog Encoding Voltage—A Key to Ultra-Wide Dynamic Range and Low Power CMOS Image Sensor, doi:10.3390/jlpea3010027

CHAPTER 12

Ultra-Low Power Designing For CMOS Sequential Circuits

Patikineti Sreenivasulu[1], Srinivasa Rao[2], Vinaya Babu[3]*

[1]ECE Department, JNTUK, Andhrapradesh, Kakinada, India
[2]ECE Department, TRR College of Engineering, Hyderabad, India
[3]CSE Department, JNTUH College of Engineering, Hyderabad, India

ABSTRACT

Power consumption is the bottleneck of system performance. Power reduction has become an important issue in digital circuit design, especially for high performance portable devices (such as cell phones, PDAs, etc.). Many power reduction techniques have also been proposed from the sys- tem level down to the circuit level. High-speed computation has thus become the expected norm from the average user, instead of being the province of the few with access to a powerful main- frame. Power must be added to the portable unit, even when power is available in non-portable applications, the issue of low-power design is becoming critical. Thus, it is evident that methodologies for the design of high-throughput, low-power digital systems are needed. Techniques for low-power operation are shown in this paper, which use the lowest possible supply voltage coupled with architectural, logic style, circuit, and technology optimizations. The threshold vol- tages of the MTCMOS devices for both low and high V_{th} are constructed as the low threshold V_{th} is approximately 150 - 200 mv whereas the high threshold V_{th} is managed by varying the thickness of the oxide T_{ox}. Hence we are using different threshold voltages with minimum voltages and hence considered this project as ultra-low power designing.

INTRODUCTION

Low power design can be exploited at various levels, e.g., system level, architecture level, circuit level, and device level [1] . This paper first gives a brief overview for low power optimization techniques at system and architecture level, reducing power consumption has become an important issue in digital circuit design [1], especially for high performance portable devices. Many power reduction techniques have also been proposed from the system level down to the circuit level. In this section, some of these techniques, which are related to the design for parallel multiplier, will be presented. MTCMOS sleep transistor sizing issues are addressed, and a hierarchical sizing methodology based on mutual exclusive discharge patterns is presented. The multi-threshold CMOS (MT-CMOS) circuit is an effective powering-down scheme in the low supply voltage range, but it needs extra circuits for the data holding in the sleep mode [2], such as the Balloon Circuits or the intermittent power supply scheme.

LOW POWER TECHNIQUES

A) Supply Voltage Scaling

To reduce the power consumption, the scaling power supply voltage is most effective method [2]. Reducing the supply voltage can significantly reduce the power dissipation that is a quadratic function of the operating voltage. This is illustrated in Figure 1, it represents the power consumption as a function of V_{dd} for a 4-bit carry look-ahead adder in 0.18 μm process technology [3] [4]. The supply voltage for various logic functions and logic styles are dependence on the power consumption.

However, the supply reducing voltage also the delay is increases. The relationship between V_{dd} and the delay, T_d, can be expressed as

$$Td = \frac{C_L \times V_{dd}}{I} = \frac{C_L \times V_{dd}}{\frac{\mu C_{ox}}{2}(W/L)(V_{dd} - V_t)^2} \quad (1)$$

From the Equation (1), when V_{dd} approaches the threshold voltage, V_t, the delay increases drastically [4], as shown in Figure 2. Obviously, using this method causes the performance loss on the speed. In order to compensate for the loss in throughput at low supply voltages, the parallel and pipelined architectures as well as modifying the threshold voltage of the devices can be applied for several techniques.

B) Reducing Effective Capacitance

When the performance loss in throughput due to lowering the supply voltage is not acceptable. The low power consumption in CMOS circuits can also effective the reducing capacitance. The effective capacitance is defined by the product of the physical capacitance and the switching activity [5], which is shown as

$$C_{\text{effective}} = \alpha_{0\text{-}1} C_L$$

where $\alpha_{0\text{-}1}$ is the node transition activity factor and C_L is the load capacitance which refers to physical capacitance. The switching power consumption can be rewritten as

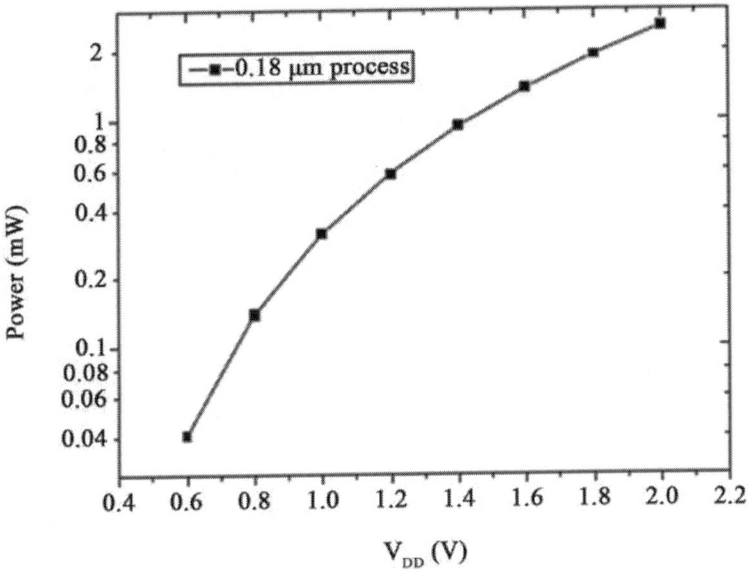

Figure 1. Power consumption for a 4-bit CLA as a function of V_{dd} [1].

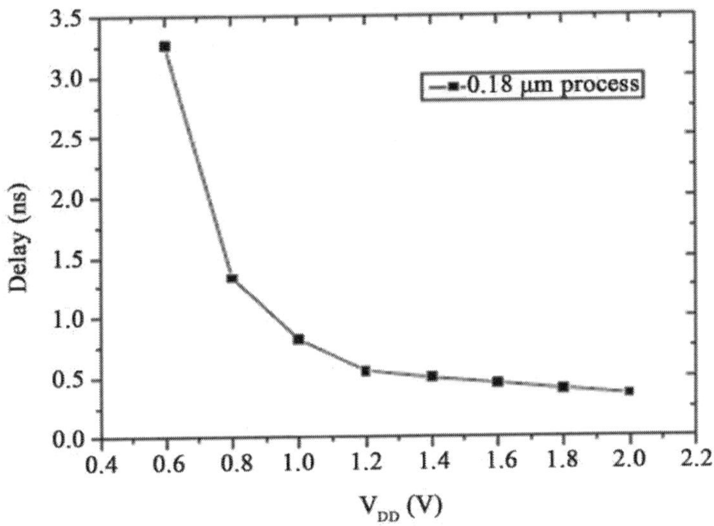

Figure 2. Propagation delay versus V_{dd} for a 4-bit CLA adder [3].

$$P_{\text{switching}} = C_{\text{effective}} V_{dd}^2 f_{clk}$$

From the above equation, reducing the switching power consumption can be achieved by minimizing both of the physical capacitance and the switching activity [6].

Physical Capacitance Reduction:

The physical capacitance can be reduced through selecting the appropriate circuit style and optimizing the transistor sizes.

Effects of Circuit Styles

The different circuit and logic styles result in different gate and diffusion capacitance of the transistors in a combinational logic circuit [6]. Some of the circuit styles can substantially reduce the physical capacitance and is good for low-power operation. Figure 3 represents the power-delay products of an 8-bit adder relationship between that was implemented in 2 µm CMOS technology with different circuit styles and the corresponding propagation delays.

As shown in Figure 2, the adder that was implemented by using complementary pass transistor logic (CPL) is about twice as fast as the conventional static CMOS. This is due to that CPL improves the performance of the circuit with a lower input capacitance and reduced voltage swing. Moreover, a CPL logic circuit consumes less power than a static CMOS one, for instance, the power saving for a CPL adder is about 30% compared to a conventional static CMOS adder [7]. This improvement is mainly due to the reduction in capacitance. The performance of a full adder implemented with different circuit styles, such as conventional CMOS, transmission gate CMOS. Power dissipation of the circuit dramatically on the delay circuit styles to reveals this comparison. The compared results indicate that the CPL-TG provides the lowest power delay product, and the LCPL2 has the second lowest power delay product. Both of them are the best suited for low-power high-performance applications such as adders and multipliers.

Transistor Sizing

The capacitive load that originates from transistor capacitance and interconnect wiring can be reduced by optimizing transistor sizes whenever possible and reasonable. In general, increasing the transistor sizes results in a large (dis)charging current and simultaneously increases the parasitic capacitance. On the other hand, reducing the transistor sizes will result in decreasing input capacitance that may be the load capacitance for other gates and lowering the speed of the circuit. Thus, the objective of transistor sizing is to obtain the minimum power dissipation under given performance requirements. In order to explain how to make transistor sizing, let us consider a static inverter driving a load capacitance being composed of an intrinsic (diffusion) and an extrinsic (wiring and fan out) capacitances. When the total load capacitance to the gate output is dominated by the diffusion capacitance, the smallest possible sizes of the transistors should be used for obtaining the lowest power consumption. Otherwise, if the load capacitance is dominated by the extrinsic component [8].

C) Switching Activity Reduction

The dynamic power consumption of a circuit is strongly related to the switching activity of the circuit [8]. At the circuit level, one main consideration for low-power designs is the choice of the static or dynamic logic styles. The dynamic logic gates are clocked, and undergo the pre-charge and evaluation phases, which are suitable for high-speed applications for the expense of high power dissipation. Whereas the static CMOS is the best choice for low-power high-speed implementation of dedicated circuit applications like multipliers .The switching activity can be reduced by many means such as reordering input signals, no bus-sharing technique, and minimizing the glitching activity of the static circuits etc.

Minimizing Glitching Activity

Glitches, or dynamic hazards, are unwanted signal transitions which occur before the signal settles to its intended value. Glitches can be generated and propagated in both data path and control parts of the circuits. The simulation result from the circuit simulator (Specter) was obtained under the different conditions. The spurious transitions consume extra power compared to the glitch-free scenarios. The number of spurious transitions in a circuit depends on the logic depth, input patterns, and intermediate carry signal states etc.

The glitching activity in static circuit designs can be minimized by selecting structures with balanced signal paths and reduced logic depth. The tree structures can be applied to implement a circuit with both of the balanced signal paths and less logic depth, while the chain structures are quite the contrary. A good example in Figure 4 illustrates the choice of the tree or chain structures [5]. The second adder computes twice and the third adder computes three times per cycle due to the finite propagation delay through the previous adders. By contrast, the logic depth in the tree case has been reduced from three to two and the signal paths are more balanced. Thus, the switched capacitance (effective capacitance) for the chained case is a factor of 1.5 larger than in the tree [8] [9].

Another possible approach to eliminate the spurious transitions is to use dynamic logic circuits instead of static logic, since any node in dynamic logic circuits can only undergo at most one transition per clock cycle.

D) Reduction Technique for Leakage Power

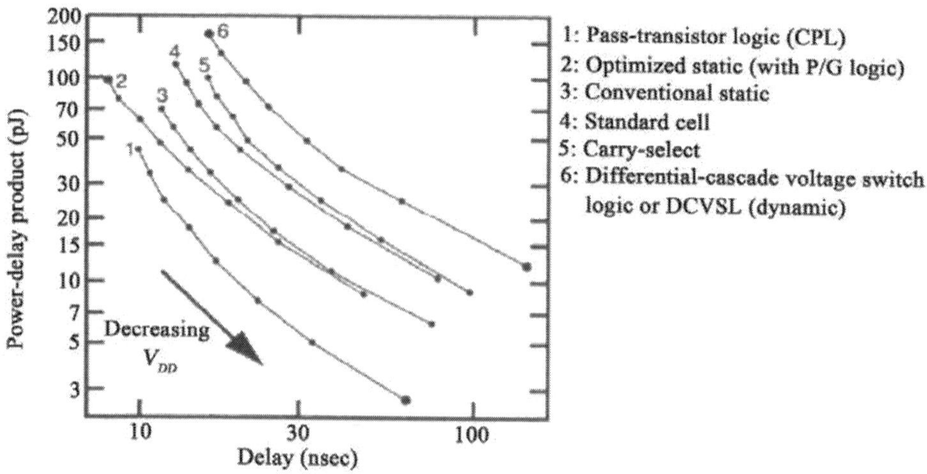

Figure 3. Power-delay products versus delay for an 8-bit adder [4].

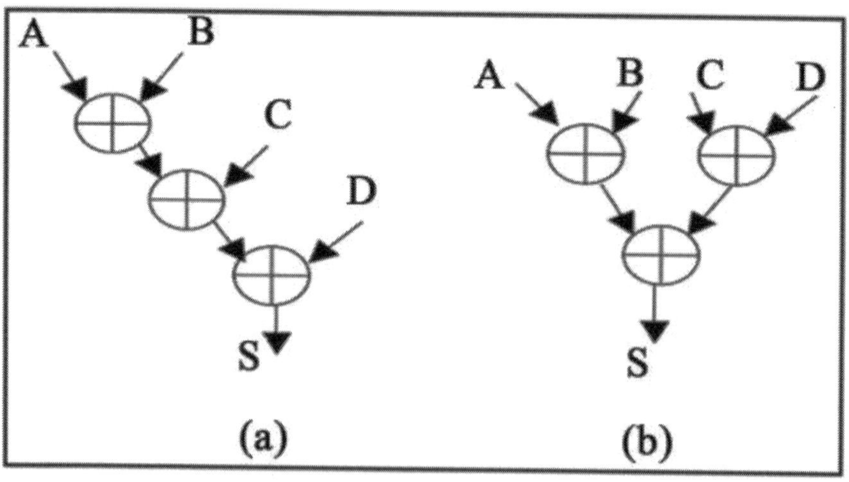

Figure 4. Tree versus chain structures [6].

The number of typical reduction for leakage techniques fall in two categories, either leveraging the stack effect or increasing the transistor threshold voltage. Forcing transistors into stacks is not a scalable solution and its implementation becomes very complicated for large designs. There are three main reasons for minimizing leakage current. They are (i) source/drain junction leakage current which is due the junction acts as an diode in reverse bias when the transistor is in sleep mode. i.e. in OFF. (ii) Is gate tunneling leakage current which flows through the oxide to substrate? If the gate oxide is thin, then the current increases exponentially. And the third one is (iii) sub threshold leakage current, which is a current from source to drain. It is diffusion current built by minority carriers in the channel in the MOS devices [9].

As technology scales into the deep-submicron (DSM) regime, standby sub threshold leakage power increases exponentially with the reduction of the supply voltage vdd and the threshold voltage vth. For many event driven applications, such as mobile devices where circuits spend most of their time in an idle state with no computation, stand by leakage power is especially detrimental on overall power dissipation [10]. Multi-Threshold CMOS (MTCMOS) is an effective circuit-level methodology that provides high performance in the active mode and saves leakage power during the standby mode. The basic principle of the MTCMOS technique is to use low vth transistors to design the logic gates where the switching speed is essential [11], while the high vth transistors (also called sleep transistors) (Figure 5) are used to effectively isolate the logic gates in standby state and limit the leakage dissipation. The sleep transistors are turned on when the circuit is active and turned off when the circuit is idle. By cutting off the power source, this technique can reduce leakage power effectively.

LEAKAGE FEEDBACK STATIC FLIP FLOP

Another use of the leakage feedback gate is to modify the static MTCMOS flip flop to eliminate the need for the parallel inverter to re circulate data. The leakage feedback structure can be used instead, which does not slow down the critical path because no extra capacitances are introduced on internal nodes [12]. The addition of the helper sleep devices only add load to the outputs of I4 and I5 which are not part of the critical path, so the speed of the MTCMOS flip flop is not compromised. When transitioning from the sleep state to the active mode state, one cannot immediately turn on the master latch high Vt sleep transistors because that might accidentally cause the data to flip state.

SIMULATION RESULTS (FIGURE 6 AND FIGURE

Simulations were performed on the various MTCMOS flip flop architectures in a 0.16 u technology with high Vt approximately 0.15 V and low V_t approximately 0.05 V (defined the 10 nA @ 10 um point). Figure 8 illustrates how the leakage feedback gate of Figure 9 can hold data during the standby state even when the input varies during the sleep state [12].

CONCLUSION

In this paper, several existing power reduction techniques are discussed and a novel circuit design technique to minimize sleep mode power consumption due to leakage power in CMOS technology is discussed. This circuit technique provides significant energy savings in sleep mode without any speed degradation or die area overhead. Moreover, it is almost independent of technology scaling and has no circuit design complexity. This paper presented several dual-threshold voltage circuit techniques that can help reduce sub threshold leakage currents during standby modes for combinational logic blocks [13]. MTCMOS was shown to be an effective standby leakage control technique for static logic, but difficult to implement since sleep transistor sizing is highly dependent on discharge patterns within the circuit block. We have developed a multi-sleep transistor synthesis

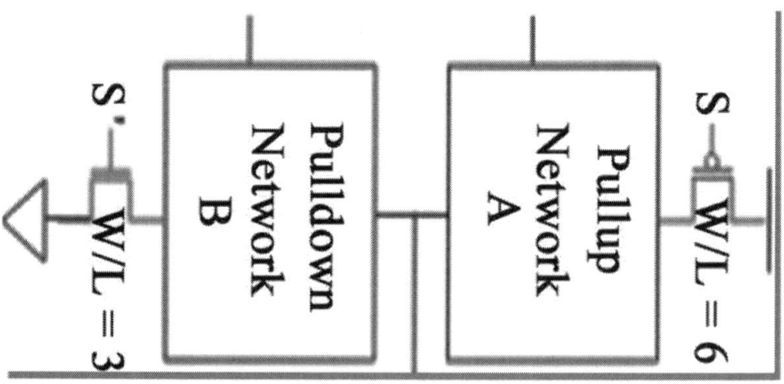

Figure 5. Sleepy transistor [8].

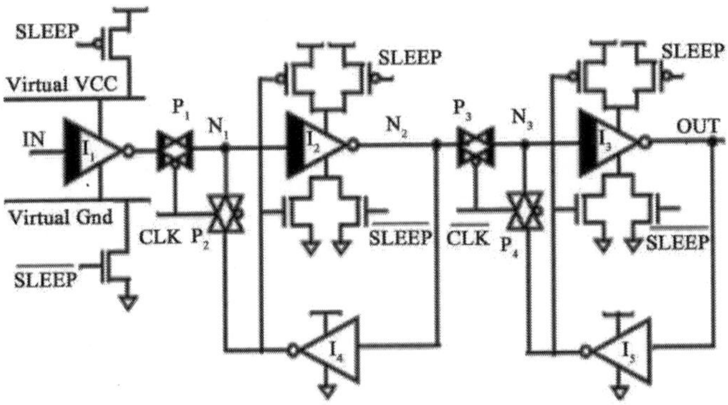

Figure 6. Leakage feedback MTCMOS flip-flop [8].

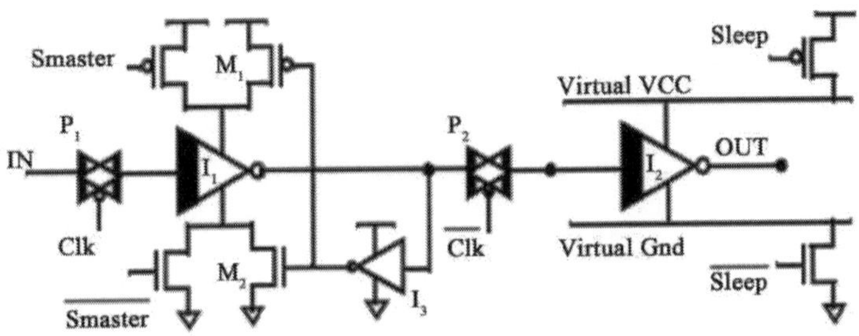

Figure 7. Dynamic flip-flop leakage feedback [9].

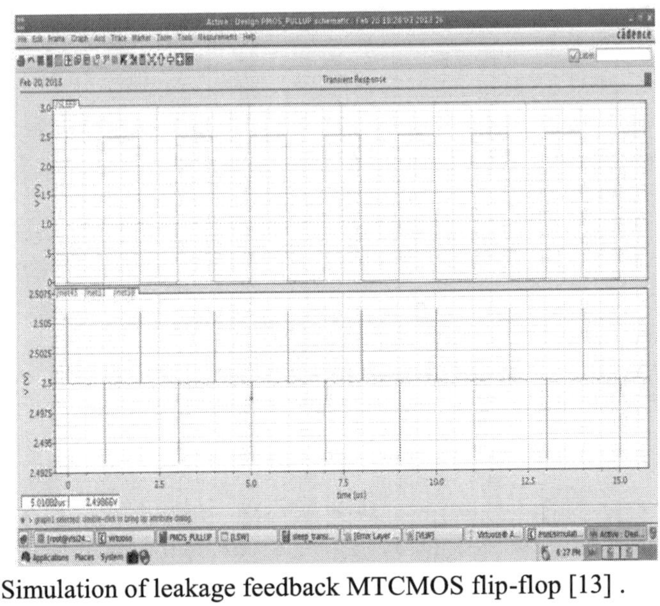

Figure 8. Simulation of leakage feedback MTCMOS flip-flop [13] .

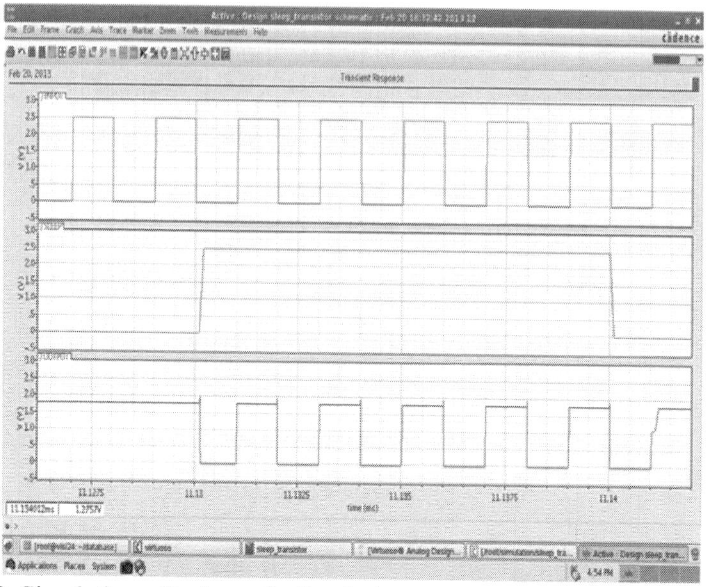

Figure 9. Simulation of leakage feedback dynamic flip-flop [13].

Technique which enables to further reduce the total leakage w.r.t. the case of high-sleep transistors [14].

REFERENCES

1. Mutoh, S., et al. (1995) 1-V Power Supply High-Speed Digital Circuit Technology with Multithreshold-Voltage CMOS. IEEE Journal of Solid-State Circuits, 30, 847-854.http://dx.doi.org/10.1109/4.400426
2. Shigematsu, S., et al. (1995) A 1-V High-Speed MTCMOS Circuit Scheme for Power-Down Applications. Symposium on VLSI Circuits, Kyoto, 8-10 June 1995, 125-126.
3. Kuroda, T., et al. (1996) A 0.9V 150MHz 10mW 4mm^2 2-D Discrete Cosine Transform Core Processor with Variable- Threshold-Voltage Scheme. IEEE International Solid-State Circuits Conference, New York, February 1996, 166-167.
4. Shimizu, T., et al. (1996) A Multimedia 32b RISC Microprocessor with 16Mb DRAM. 1996 IEEE International Solid-State Circuits Conference, San Jose, February 1996, 216-217.
5. Kumagai, K., Iwaki, H., Yoshida, H., Suzuki, H., Yamada, T. and Kurosawa, S. (1998) A Novel Powering-Down Scheme for Low Vt CMOS Circuits. Symposium on VLSI Circuits, Honolulu, 11-13 June 1998, 44-45.

6. Makino, H., Tsujihashi, Y., Nii, K., Morishima, C., Hayakawa, Y., Shimizu, T. and Arakwa, A. (1998) An Auto- Backgate-Controlled MT-CMOS Circuit. Symposium on VLSI Circuits, Honolulu, 11-13 June 1998, 42-43.
7. Brodersen, R.W., Chandrakasan, A. and Sheng, S. (1993) Design Techniques for Portable Systems. IEEE International Solid-State Circuits Conference, San Jose, February 1993, 168-169.
8. Chandrakasan, A.P., Sheng, S. and Brodersen, R.W. (1992) Low-Power CMOS Digital Design. IEEE Journal of Solid- State Circuits, 27, 473-484.http://dx.doi.org/10.1109/4.126534
9. Horiguchi, M., Sakata, T. and Itoh, K. (1993) Switched-Source-Impedance CMOS Circuit for Low Standby Subthreshold Current Giga-Scale LSI's. IEEE Journal of Solid-State Circuits, 28, 1131-1135. http://dx.doi.org/10.1109/4.245593
10. Kawahara, T., Horiguchi, M., Kawajiri, Y., Kitsukawa, G., Kure, T. and Aoki, M. (1993) Subthreshold Current Reduction for Decoded-Driver by SelfReverse Biasing. IEEE Journal of Solid-State Circuits, 28, 1136-1144. http://dx.doi.org/10.1109/4.245594
11. Mutoh, S., Douseki, T., Matsuya, Y., Aoki, T. and Yamada, L. (1993) 1V HighSpeed Digital Circuit Technology with 0.5-μm Multi Threshold CMOS. Proceedings of 6th Annual IEEE International ASIC Conference and Exhibit, Rochester, 27 September-1 October 1993, 186-189.
12. Sreenivasulu, P., Srinivasa Rao, K. and Vinayababu, A. (2011) Power Optimization in Digital Circuits through Leakage Current Reduction by Using Multi Threshold CMOS. International Journal of VLSI Signal Processing Applications, 1, No. 5.
13. Kao, J.T. and Chandrakasan, A.P. (2000) Dual-Threshold Voltage Techniques for Low-Power Digital Circuits. IEEE Journal of Solid-State Circuits, 35, 1009-1018.
14. Shigematsu, S., et al. (1997) A 1-V High Speed MTCMOS Circuit Scheme for Power-DOWN Application Circuits. IEEE Journal of Solid-State Circuits, 32, 861-869

CITATION

Sreenivasulu, P., Rao, S. and Babu, V. (2015) Ultra-Low Power Designing for CMOS Sequential Circuits. International Journal of Communications, Network and System Sciences, 8, 146-153. doi: 10.4236/ijcns.2015.85016.

CHAPTER 13

A Low Power CMOS Imaging System with Smart Image Capture and Adaptive Complexity 2D-DCT Calculation

Qing Gao and Orly Yadid-Pecht

Department of Electrical and Computer Engineering, University of Calgary, AB T2N1N4, Canada;

ABSTRACT

A novel low power CMOS imaging system with smart image capture and adaptive complexity 2D-Discrete Cosine Transform (DCT) is proposed. Compared with the existing imaging systems, it involves the smart image capture and image processing stages cooperating together and is very efficient. The type of each 8 × 8 block is determined during the image capture stage, and then input into the DCT block, along with the pixel values. The 2D-DCT calculation has adaptive computation complexity according to block types. Since the block type prediction has been moved to the front end, no extra time or calculation is needed during image processing or image capturing for prediction. The image sensor with block type decision circuit is implemented in TSMC 0.18 μm CMOS technology. The adaptive complexity 2D-DCT compression is implemented based on Cyclone EP1C20F400C8 device. The performance including the image quality of the reconstructed picture and the power consumption of the imaging system are compared to those of traditional CMOS imaging systems to show the benefit of the proposed low power algorithm. According to simulation, up to 46% of power consumption can be saved during 2D DCT calculation without extra loss of image quality for the reconstructed pictures compared with the conventional compression methods.

INTRODUCTION

Wide utilization of portable battery-operated devices in multimedia applications, such as cell phones, portable digital assistants (PDAs) and smart toys, has triggered a demand for ultra low-power image system. CMOS imaging technology has recently become a very attractive solution for these applications as they consume less power, and operate at higher speeds compared with CCD imaging technology [1].

Many low power designs for image capture were reported during the last decade [2,3,4,5,6,7,8]. A review of low power designs in CMOS image sensors at different levels is given in [2]. Some works aim at compensating the reduced signal to noise ratio and dynamic range caused by a low operating voltage [3]. In [4], SOI (Silicon-On-Insulator) technology is used instead of the traditional CMOS technology since it has smaller parasitic capacitance and reduced leakage current. In 2006, the first self-powered image sensor was proposed by Fish *et al.* [5]. Then an optimized energy harvesting CMOS image sensor was proposed [6], where the photo detector itself can be used for power generation besides the PGPd. In [7, 8] a block based dual VDD image sensor was proposed. It has dual supply voltages during image capture stage and the supply voltage is decided according to the block types.

After the image capture, energy-aware data compression is usually performed for efficient transmission. The Discrete Cosine Transform (DCT), and in particular the DCT-II, is often used in image/video processing such as JPEG still image compression, MJPEG, MPEG video compression due to its good energy compaction. However the DCT itself contains very computationally intensive matrix multiplications and therefore is power consuming. Numerous algorithms have been proposed attempting to minimize the number of additions and multiplications such as the Loeffler DCT [9,10,11] or even to replace multiplications with only add and shift operations, *i.e.*, Distributed Arithmetic (DA), Coordinate Rotation Digital Computer (CORDIC) and bin DCT [12,13,14]. Also, data-dependent DCT algorithms have been introduced for low power purpose [15].

In all the existing CMOS imaging systems, the image capture stage and the compression stage are simply concatenated together. The DCT architectures without multiplier are usually time consuming and hardware-expensive. Some DCT designs subsample the area where pixels change less, but all the block type predictions are made during the digital image processing and require extra processing time and also extra memory space to store the image data during prediction. However, in our imaging system, these two stages are intertwined together. The block type prediction is moved to the front end, that is, the block type is estimated during image capture in an analog way (at the mean time during read out). It is more efficient in terms of power. In addition, no extra time is needed during image processing or image capturing for block type prediction. The 2D DCT calculation has adaptive data format and computation complexity depending on the block type.

The paper is organized as follows. In Section 2, the general algorithm and architecture of the proposed low power imaging system is described. Section 3discusses the circuit implementation of the imager with the block decision circuit and the adaptive-complexity 2D-DCT. Performance regarding the image quality and the power consumption will be also given in Section 3. Finally, conclusions are presented in Section 4.

PROPOSED LOW POWER IMAGING SYSTEM WITH SMART IMAGE CAPTURE AND ADAPTIVE COMPLEXITY 2D-DCT

Traditional CMOS Imaging System with 2D-DCT Calculation

In a traditional CMOS imaging system, the image is first captured by the imager, and then the analog image data go through an Analog-to-Digital (ADC) conversion and then digital processing, such as image compression. The general diagram of a basic traditional digital camera system with 2D-DCT based compression is shown in Figure 1.

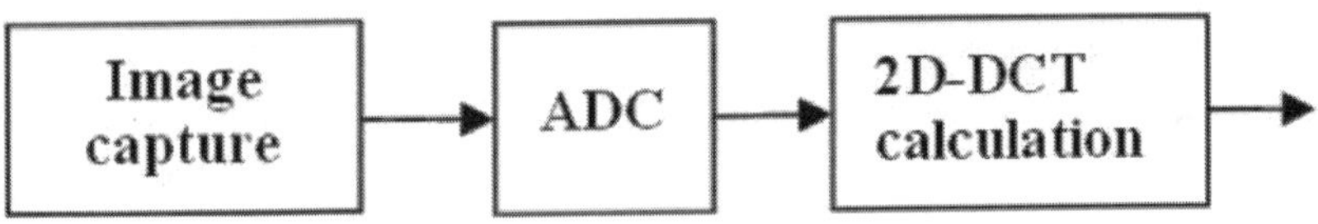

Figure 1. A traditional digital imaging system with 2D-Discrete Cosine Transform (DCT) calculation.

In hardware the N × N 2D-DCT can be realized by storing the output of the first 1-D DCT in a memory buffer line after line, then applying a second 1-D DCT transform on the columns of the results [16]. N is typically 8 in most of the applications, resulting in an 8 × 8 transform coefficient.

Proposed Low Power Imaging System with Smart Image Capture and Adaptive Complexity 2D-DCT Calculation

Architecture of the Proposed Low Power Imaging System

The architecture of the novel low power digital camera system with adaptive complexity 2D-DCT based calculation is shown in Figure 2.

The system is mainly composed of an image sensor for smart image capture, an ADC for data format converting and an adaptive-complexity 2D-DCT calculation. In addition to capturing images as the traditional CMOS imagers, the image sensor in the proposed camera system contains a block type decision block that can compute the type for each block of 8 × 8 pixels according to the variance [17]. In order to simplify the implementation, we use the difference between the maximum and minimum values within a block as an approximation of the variance to represent how far a set of numbers is spread out, similar to what was done in [8]. A block with lower $Vmax - Vmin$ has a trend of lower variance. This is more accurate when the variance is small, so we use $Vmax - Vmin$ as an approximation of the variance to simplify the implementation. The complexity of the proposed 2D DCT calculations is dependent on the block type decided in the image sensor during image capture stage. Power saving is achieved with reduced computation during DCT for small variance blocks.

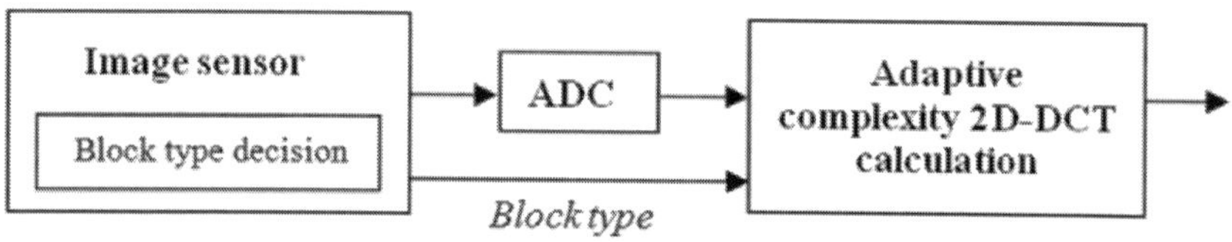

Figure 2. The proposed low power digital CMOS imaging system with smart image capture and adaptive complexity 2D-DCT calculation.

Adaptive Complexity Compression

As shown in Figure 3, for blocks with large $Vmax - Vmin$ (object blocks), conventional DCT is performed. For the blocks with small $Vmax - Vmin$ (background blocks), the differential data format is used instead of the pixel value itself during AC coefficients calculation. In addition, the resolution unit is considered as N × N to reduce the computation complexity. Here N can be selected from 1, 2, 4 and 8 according to different applications. In our implementation, N is chosen as 2 for now. With reduced spatial resolution, part of the computation can be skipped without loss of useful information. Because less bits are used during the AC coefficients computing and part of the calculation is skipped, power saving is achieved for background blocks.

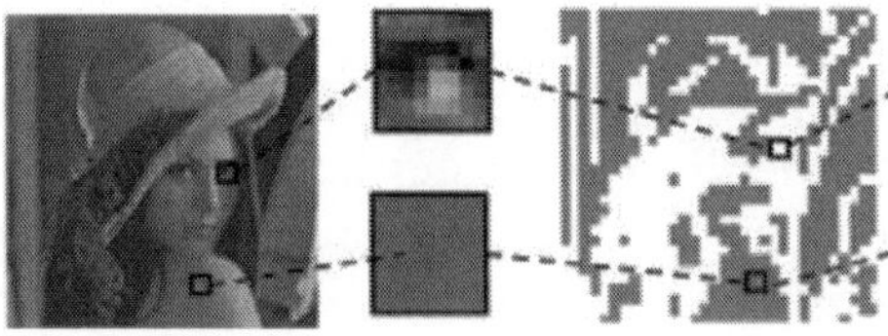

Figure 3. Example of the proposed low power algorithm.

Using vector form, the 8 × 8 DCT transform becomes $F = CXC^T$. Where X is the input matrix and F is the output results after DCT transform. C is the cosine coefficients matrix and C^T is the transpose coefficients matrix.

For background type blocks, the spatial resolution unit will be considered as 2 × 2 during calculation, therefore the row DCT transform becomes $F' = CX'C^T$, where $C$is the new cosine coefficients and C^T is the new transpose coefficients. X' is the new pixel block and F' is the corresponding result after DCT transform.

$$\begin{aligned} F' &= CX'C^T = C_{8x8}X'_{8x8}C_{8x8}{}^T = C_{8x8}(B_{8x4}X''_{4x4}B_{8x4}{}^T)C_{8x8}{}^T \\ &= (C'_{8x4})X''_{4x4}(C'_{8x4})^T \end{aligned} \quad (1)$$

Where

$$B_{8x4} = \begin{bmatrix} 1 & 1 & 0 & 0 & 0 & 0 & 0 & 0 \\ 0 & 0 & 1 & 1 & 0 & 0 & 0 & 0 \\ 0 & 0 & 0 & 0 & 1 & 1 & 0 & 0 \\ 0 & 0 & 0 & 0 & 0 & 0 & 1 & 1 \end{bmatrix}^T, X_{4x4}'' = \begin{bmatrix} x0 & x2 & x4 & x6 \\ x16 & x18 & x20 & x22 \\ x32 & x34 & x36 & x38 \\ x48 & x50 & x52 & x54 \end{bmatrix},$$

$C'_{8x4} = C_{8x8}B_{8x4}$, and the subscriptions indicate the sizes of the matrixes. So the matrix sizes during computation can be reduced and the times of multiplication and addition are also reduced. In hardware, it is realized by skipping part of the calculations for background type blocks. Similarly, during column DCT, part of the calculation can also be skipped since half of the inputs in each column are the same.

The analog data are read out from imager row by row. After ADC, the data needs to be stored in a memory temporarily for reordering before performing DCT. If the block type prediction is done during image processing, since the digital data is read out from the memory one by one, at least 64 additional clocks are needed in order to make a prediction for an 8 × 8 block. Also, additional memory is required to keep the data during prediction before DCT is performed. However, in our case, since the block type is decided in the imager during the image capture stage rather than during the image processing stage, it does not need additional processing time. Also the idea of adaptive complexity 2D-DCT can be combined with the dual analog Vdd algorithm proposed in [8], which makes the best of the block type decision circuitry.

IMPLEMENTATION OF THE LOW POWER IMAGING SYSTEM AND RESULTS

Image Sensor for Smart Image Capture with Block Type Decision Block

The proposed imager for smart image capture is mainly composed of a pixel array, row and column decoders, block decision block, and readout circuits, as shown in Figure 4. It is similar to the imager proposed in [8] but not exactly the same.

It works in a rolling shutter mode, the signals do not need to be sampled to in-pixel capacitors as required in [8] but to column sample and hold circuit. The readout and the decision operations share the same row select signals and row select transistors. However in [8], the imager works in a global shutter way and the decisions are made at the middle of the integration time, therefore the read out and the decision have separate row select signals and row select transistors. The pixel circuitry here is much simpler than that in [8].

The conventional 3T CMOS APS pixel is used in the pixel array [1]. A p-channel source follower is used to compensate for threshold voltage level-shifting from the n-channel, pixel-level source follower.

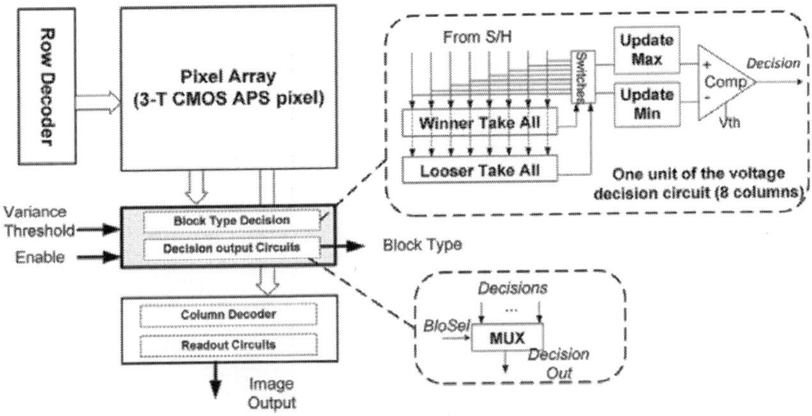

Figure 4. Block diagram of the proposed imager for smart image capture.

The block type decision unit is shared by each 8 columns and computes the type for each block according to the estimated voltage variance values V_{Max}-V_{Min}, similar to what was done in [8]. The enable signal activates the computations only when needed to save power. The block type decision unit is shown in Figure 4. Detailed description about Winner Take All (WTA), Loser Take All (LTA), Update Max/Min circuitries can be found in [8]. During the readout, the decision signal for each 8 × 8 block is output through a multiplexer one by one to the compression module for adaptive complexity controlling.

A chip for image capture and block type prediction is implemented based on TSMC 0.18 μm process, as shown in Figure 5. Its attributes are given in Table 1. The simulations are done by Cadence Spectre.

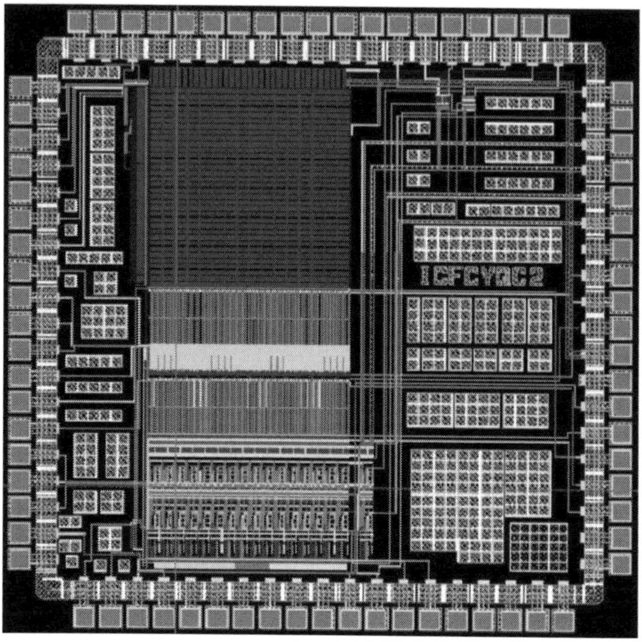

Figure 5. Layout of the image sensor with block type decision circuitry.

Table 1. Chip Attributes.

Technology	TSMC 0.18 μm
Voltage supply	1.8 V
Pixel array size	128 × 128
Pitch width	5 μm
Chip size	2 mm × 2 mm
Fill factor	26%
Estimated power (whole chip)	0.5 mW @ 30 FPS
Estimated power (decision logic)	7 μW @ 30 FPS

Image Sensor for Smart Image Capture with Block Type Decision Block Adaptive Complexity 2D-DCT Calculation

2D DCT can be done by running a 1D DCT over every row and then every column. Vector processing using parallel multipliers is a method used for implementation of DCT. The advantages of vector processing method are regular structure, simple control and interconnect and good balance between performance and complexity of implementation. The complexity of the 2D-DCT depends on the block type decided by the image sensor during image capture. Two optimizations are performed for the small variance blocks to save power.

Adaptive Data Format

For small variance blocks, only the differential part of the pixel values V_{pixel}-$V_{DC}$are used for AC coefficients computing. Here, the VDC is the minimum value in each corresponding 8 × 8 block. In order to simplify the implementation and reduce the hardware requirement, the first pixel is used as the DC part instead of the minimum pixel in the block.

Figure 6 shows the circuit for implementing the adaptive input format according to block types. For large variance blocks, the pixel values are put for DCT calculation directly. For small variance blocks, the inputs are calculated by subtracting the first pixel value of a block V_{first_b} from the pixel values, and then input to the DCT block. The DC values should be compensated to the DC coefficients to generate the final DC coefficients. Another benefit brought by this optimization is a small increment in image quality of the reconstructed picture. The reason is that because fewer bits are performed for DCT coefficients calculation, less information is lost during the truncation stage. There are less values toggling during DCT coefficient calculation and therefore less power is consumed.

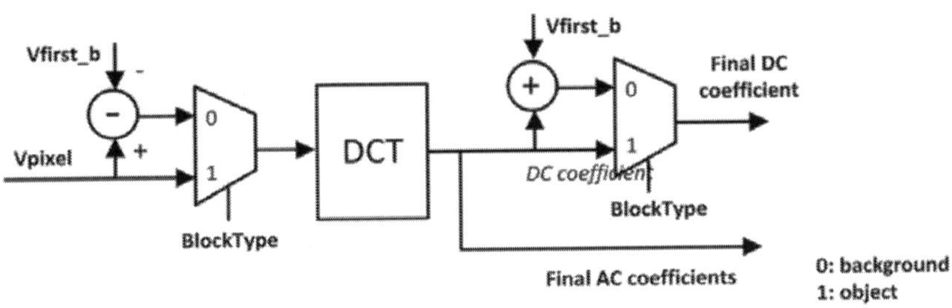

Figure 6. Schematic of adaptive input format according to block types.

Adaptive Spatial Resolution

For small variance blocks, the spatial resolution for these blocks can be reduced while not affecting the image quality much. Consequently part of the calculations can be skipped to save power consumption during DCT. For small variance blocks, the calculation for the second row is just the same as the first row, therefore we can skip the row DCT alternatively. In addition, since half of the inputs of the non-skipped row DCT and column DCT are the same, half of the calculation groups can be skipped during calculation for small variance blocks, as shown in Figure 7.

For now, the adaptive complexity 2D-DCT is implemented based on FPGA (Cyclone EP1C20F400C8) first. Later we are planning to integrate the imager for capture and the compression on the same chip. According to the synthesis report given by Quartus II, there is about 10% hardware increase than a conventional 2D-DCT with unique complexity. The maximum frequency is 100 MHz.

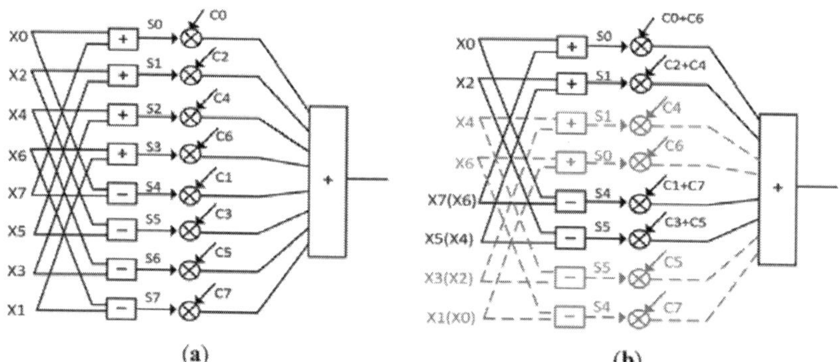

Figure 7. Schematic of adaptive spatial resolution according to block types. (a) Normal case—large variance blocks; (b) Low-complexity case (dashed paths are disabled)—Small variance blocks.

Performance

In order to show the benefit of the proposed low power algorithm, the proposed 2D-DCT based computation and a reference 2D-DCT core released by Xilinx [16] are implemented and compared.

As shown in Figure 8, three images of "Camera man", "Plane" and "Garden" which have background ratio of 50%, 90% and 0.8% small variance block ratios are used to represent three different types of images. The images have 8 bits resolution.

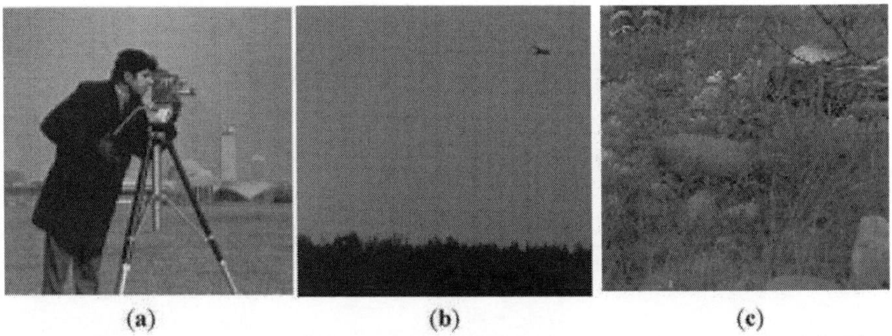

Figure 8. Test images (**a**) Camera man (background ratio is 50%); (**b**) Plane (background ratio is 90%); (**c**) Garden (background ratio is 0.8%).

Simulations about the PSNR *vs.* Variance threshold at different quantization levels are given in Figure 9a. At higher quantization level, the PSNR degradation is smaller. Therefore our low power algorithm is more efficient at higher quantization levels. The worst case happens when there is no quantization performed. The relationship between the Compression Ratio (CR) and variance threshold is given in Figure 9b. Since we have not applied entropy encoding yet, the Compression Ratio (CR) here is expressed by level of the quantization, that is, the percentage of the non-zero coefficients after quantization. The change of the compression ratio is not big since the compression is mainly done by DCT and quantization, the sub sampling by 2 × 2 on background blocks does not add much to the compression ratio. How PSNR changes with the variance threshold also depends on image types. Figure 10 shows the reconstructed image quality and power consumption for different types of images at different variance thresholds compared with those of a traditional compression in worst case (no quantization is performed).

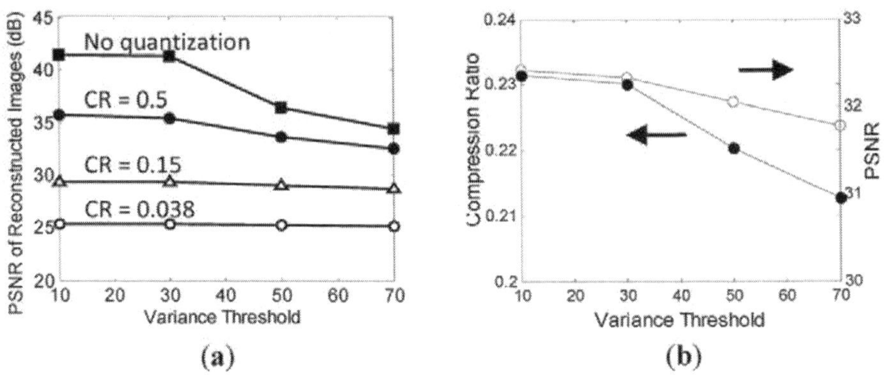

Figure 9. (**a**) PSNR of reconstructed images *vs.* Variance threshold at different quantization levels; (**b**) Compression ratio and PSNR *vs.* Variance threshold.

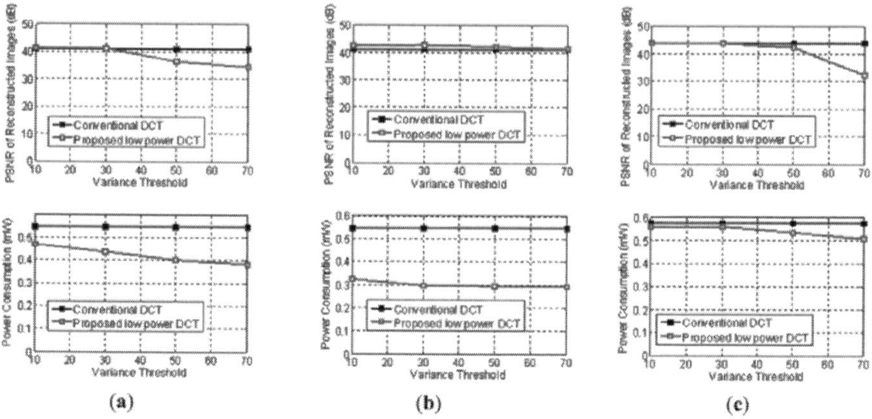

Figure 10: (**a**) PSNR and Power *vs*. Variance threshold for normal background images—Cameraman; (**b**) PSNR and Power *vs*. Variance threshold for flat background images—Plane; (**c**) PSNR and Power *vs*. Variance threshold for busy background images—Garden.

The power is estimated by Quartus II based on the Voltage Change Dump (VCD) files from post layout netlist simulation and the PSNR analysis is done by Matlab. The clock frequency used here is 0.5 MHz corresponding to a 128 × 128 array working at 30 FPS. It can be increased up to 100 MHz for imagers with larger array size and higher frame rate. The power savings varies from image to image. It is more efficient for predominant background images with up to about 46% power saving while no extra image quality degradation is observed compared with traditional compression. Extra image quality degradation is small because optimizations are performed only for background blocks. Power saving and the image quality of the reconstructed picture depend on the variance threshold. It is a tradeoff between image quality and power, and can be easily controlled by the threshold according to different applications. As shown in Table 2, by choosing the appropriate variance threshold, *i.e.*, 30 out of 256 for 8-bit resolution images, significant amounts of power can be saved with no extra image quality degradation.

Table 2. Power saved During 2D-DCT Calculation.

Images	Background ratio (th = 30)	Percentage of power saving	Extra image quality degradation
Garden	0.8%	0.5%	None
Cameraman	50%	24%	None
Plane	90%	46%	None

The power of the whole imaging system is the sum of the image sensor, ADC and the compression. For the proposed image sensor, the power estimation of the block type prediction is 7.0 μW. This adds only about 1.4% to the total imager power, and 0.7% to the system. Therefore, we conclude that the expected power savings outperform the extra power caused by the block type decision circuitry and result in significant power savings for the system.

CONCLUSIONS

A novel low power CMOS imaging system with smart image capture and adaptive complexity 2D-DCT calculation is proposed, simulated and implemented. The complexity of the 2D-DCT calculation is controlled by the block types which are estimated during image capture stage. It does not add additional processing time or memory space for block type prediction.

The imager is more efficient when the picture has predominant background. By choosing appropriate threshold, up to 46% of the power consumption can be saved during 2D-DCT calculation for images having predominant background, while no extra image quality degradation occurs for the reconstructed pictures compared with traditional compressions. For typical scenarios, up to about 23% of power can be saved for the whole imaging system. The idea of smart image capture and adaptive complexity according to block types can be extended to other 2D DCT architectures.

ACKNOWLEDGMENTS

The authors would like to thank CMC for the support with Cadence tools and chip fabrication.

REFERENCES

1. Yadid-Pecht, O.; Etienne-Cummings, R. *CMOS Imagers: From Phototransduction to Image Processing*; Kluwer: Norwell, MA, USA, 2004.
2. Fish, A.; Yadid-Pecht, O. Low-Power "Smart" CMOS Image Sensors. In Proceedings of the IEEE International Symposium on Circuits and Systems, Washington DC, USA, 18–21 May 2008; pp. 1408–1411.
3. Xu, C.; Zhang, W.; Ki, W. A 1.0 V VDD CMOS active-pixel sensor with complementary pixel architecture and pulsewidth modulation fabricated with a 0.25 μm CMOS process. *IEEE J. Solid-State Circuits* **2002**, *37*, 1853–1859.
4. Shen, C.; Xu, C.; Huang, R.; Zhang, W.; Ko, P.K.; Chan, M. A New APS Architecture on SOI Substrate for Low Voltage Operation. In Proceedings of the 9th International Symposium on IC Technology, Systems & Applications, Singapore, 3–5 September 2001; pp. 275–278.
5. Fish, A.; Hamami, S.; Yadid-Pecht, O. CMOS image sensors with self-powered generation capability. *IEEE Trans. Circuits Syst. II* **2006**, *53*, 131–135.
6. Shi, C.; Law, M.K.; Bermak, A. A novel asynchronous pixel for an energy harvesting CMOS image sensor. *IEEE Trans. VLSI Syst.* **2011**, *19*, 118–129.
7. Gao, Q.; Yadid-Pecht, O. Dual VDD Block Based CMOS Image Sensor–Preliminary Evaluation. In Proceedings of the IEEE International Symposium on Circuits and Systems (ISCAS), Rio de Janeiro, Brazil, 15–19 May 2011; pp. 1820–1823.
8. Gao, Q.; Yadid-Pecht, O. A low power block based CMOS image sensor with dual VDD. *IEEE Sens. J.* **2012**, *12*, 747–755.
9. Loeffler, C.; Lightenberg, A.; Moschytz, G.S. Practical Fast 1-D DCT Algorithms with 11-Multiplications. In Proceedings of the International Conference on AcousticsSpeechand Signal Processing-89, Glasgow, UK, 23–26 May 1989; Volume 2, pp. 988–991.

10. Thoudam, V.P.S.; Bhaumik, B.; Chatterjee, S. Ultra Low Power Implementation of 2-D DCT for Image/Video Compression. In Proceedings of International Conference on Computer Applications & Industrial Electronics (ICCAIE 2010), Kuala Lumpur, Malaysia, 5–7 December 2010; pp. 532–536.
11. Sun, C.-C.; Ruan, S.-J.; Heyne, B.; Goetze, J. Low-power and high-quality Cordic-based Loeffler DCT for signal processing. *Circuits Devices Syst. IET* **2007**, *1*, 453–461.
12. Tran, T.D. The binDCT: Fast multiplierless approximation of the DCT. *IEEE Signal Process. Lett.* **2000**, *7*, 141–144.
13. Sung, T.-Y.; Shieh, Y.-S.; Yu, C.-W.; Hsin, H.-C. High-Efficiency and Low-Power Architectures for 2-D DCT and IDCT Based on CORDIC Rotation. In Proceedings of the 7th International Conference on Parallel and Distributed Computing Applications and Technologies, Taipei, Taiwan, 4–7 December 2006; pp. 191–196.
14. Jeong, H.; Kim, J.; Cho, W.-K. Low-power multiplierless DCT architecture using image data correlation. *IEEE Trans. Consum. Electron.* **2004**, *50*, 262–267.
15. Xanthopoulos, T.; Chandrakasan, A.P. A low-power DCT core using adaptive bitwidth and arithmetic activity exploiting signal correlations and quantization. *IEEE J. Solid-State Circuits* **2000**, *35*, 740–750.
16. Pillai, L. *Video Compression Using DCT*; Xilinx: San Jose, CA, USA, 2002.
17. Loeve, M. *Probability Theory, Graduate Texts in Mathematics*, 4th ed.; Springer-Verlag: Berlin, Germany, 1977.

CHAPTER 14

CMOS Compatible Bulk Micromachining

John Ojur Dennis, Farooq Ahmad and M. Haris Khir

Department of Fundamental and Applied Sciences, Department of Electrical and Electronic Engineering, UniversitiTeknologi PETRONAS, Perak, Malaysia

INTRODUCTION

The last two decades have seen the emergence and prevalence of Micro-Electro-Mechanical Systems (MEMS) technology. The reduction in size, low power consumption and low cost is the ultimate goals of application specific MEMS devices. One way to achieve these goals is the monolithic integration of MEMS technology with the standard integrated circuit (IC) technology. The utilization of the conventional IC technology as a platform for design of the fast-growing MEMS technology has led to development of numerous devices and technologies in the past years. Much effort has been made and large capital has been invested into mainstream Complementary Metal-Oxide Semiconductor (CMOS) compatible MEMS technology. This approach has enabled MEMS devices to be directly integrated with CMOS circuits, allowing smaller device sizes and higher performances. This integration of MEMS technology with the mainstream CMOS technology is referred to as CMOS-MEMS technology [1].

BULK MICROMACHINING TECHNOLOGIES

Among all micromachining technologies, Bulk micromachining is the oldest one. To make small mechanical components using bulk micromachining technology, substrate material such as silicon wafer is selectively etched. There are two major classes of bulk micromachining: wet bulk micromachining (WBM) and dry bulk micromachining (DBM).WBM gives nice feature resolution and vertical sidewalls for deep trenches in the substrate but have numerous disadvantages such as incompatibility with microelectronic circuits and higher cost due to mask requirement. The fabrication cost, time and simplification of micromachining for MEMS devices is reduced by using dedicated MEMS foundry and making the prototyping development cycle much shorter by reducing number of masks and photolithography steps. CMOS MEMS technology is a good solution for post micromachining because in this technology CMOS circuitry is protected completely by the top metal layer and the same top metal layer is used as mask to define MEMS structures in MEMS region during the post-CMOS micromachining. That is why this post micromachining process is known as CMOS-compatible.

Wet Bulk Micromachining

Wet bulk micromachining is done normally with the help of chemicals. In chemical wet etching, the silicon substrate is immersed into the solution of reactive chemicals which etches the exposed regions of substrate at measureable rates. Wet etching typically uses alkaline liquid solvents, such as potassium hydroxide (KOH) or tetramethylammonium hydroxide (TMAH) to dissolve silicon which has been left exposed by the photolithography masking step. These alkali solvents dissolve the silicon in a highly anisotropic way, with some crystallographic orientations dissolving up to 1000 times faster than others. Fig. 1 shows the three major crystal planes in a cubic unit of silicon.

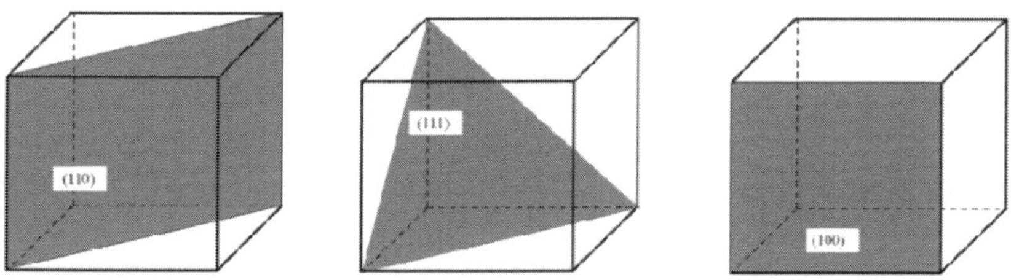

Figure 1. The three orientations <110>, <111> and <100> are the respective directions normal to the plane

For example in silicon anisotropic etching, planes <111> are etched at slower rate than all other planes. The reasons for the slow etch rate of planes <111> are, high density of silicon atoms exposed to the etchant solution in this direction and three silicon bonds laying below the plane. Fig. 2 (a) shows a schematic and (b) three dimensional photo of a typical wet anisotropic etching of a silicon substrate.

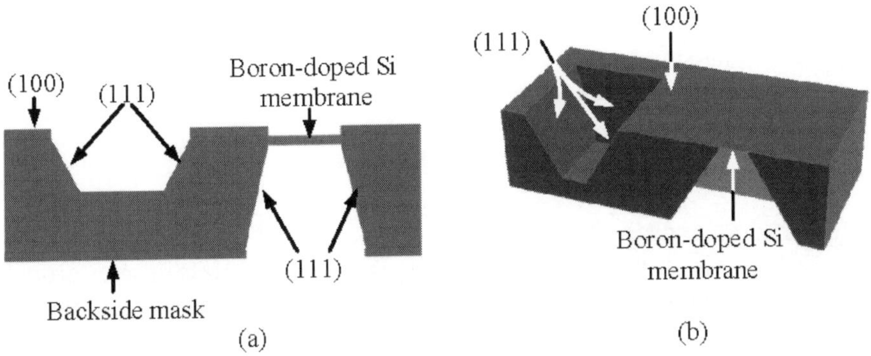

Figure 2. a) 2-D Schematic and (b) 3-D diagram of a typical wet bulk micromachining of Si

Wet chemical micromachining technology is popular in MEMS industry because of high etch rate and selectivity. There is one drawback and that is the mask is also etched during normal etching process so it is suggested to find a mask that doesn't dissolve or at least dissolves at a much slower rate than the silicon substrate. In wet etching, etch rates and selectivity can be modified by various methods such as: (a) chemical composition of etching solution, (b) dopant concentration in the substrate, (c) adjusting temperature of etching solution and finally (d) Modifying crystallographic planes of the substrate.

Chemical wet etching in bulk micromachining may be further subdivided into two parts: isotropic wet etching where the etch rate is not dependent on crystallographic orientation of the substrate and etching proceeds in all directions at equal rates as shown in Fig.3 (a) and anisotropic wet etching where the etch rate is dependent on crystallographic orientation of substrate as shown in Fig. 3 (b). Often in order to control etching process and uniform etch depths across the wafer, etch stops are used. Generally three types of etch stops are used in micromachining: Dopant etch stops, Electrochemical etch stops and Dielectric etch stop.

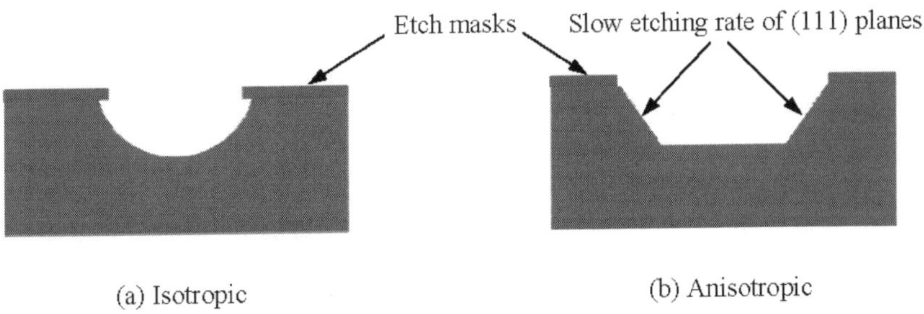

Figure 3. Difference between (a) isotropic and (b) anisotropic wet etching.

Usually in order to control etching process and uniform etch depths across the wafer, etch stops are used. Generally three types of etch stops are used in micromachining: Dopant etch stops, Electrochemical etch stops and Dielectric etch stop [2].

Dry Bulk Micromachining

Dry etching in bulk micromachining is classified into three groups: Reactive ion etching (RIE), Vapor phase etching and Sputter etching. Reactive ion etching (RIE) uses both physical and chemical mechanisms to achieve high levels of resolution. The process is one of the most diverse and most widely used processes in industry and research. Since the process combines both physical and chemical interactions, the process is much faster. The high energy collision from the ionization helps to dissociate the etchant molecules into more reactive species. In the RIE-process, cations are produced from reactive gases which are accelerated with high energy to the substrate and chemically react with the silicon. The typical RIE gasses for Si are CF_4, SF_6 and $BCl_2 + Cl_2$.

Both physical and chemical reaction is taking place with the physical part similar to the sputtering deposition. If the ions have high enough energy, they can knock atoms out of the material to be etched without a chemical reaction. It is very complex task to develop dry etch processes that balance chemical and physical etching, since there are many parameters to adjust. By changing the balance it is possible to influence the anisotropy of the etching with the chemical part being isotropic and the physical part highly anisotropic and therefore the combination can form side walls that have shapes from rounded to vertical.

A special subclass of RIE which continues to grow rapidly in popularity is deep RIE (DRIE). In this process, etch depths of hundreds of microns can be achieved with almost vertical sidewalls. The primary technology is based on the so-called "Bosch process", named after the German company Robert Bosch which filed the original patent, where two different gas compositions are alternated in the reactor. The first gas composition etches the substrate, and the second gas composition creates a polymer on the surface of the substrate. The polymer is immediately sputtered away by the physical part of the etching, but only on the horizontal surfaces and not the sidewalls. Since the polymer only dissolves very slowly in the chemical part of the etching, it builds up on the sidewalls and protects them from etching and working as pasivation, as illustrated in fig. 4. In fig.4 (a) SF_6 is etching the silicon while in fig. 4 (b) C_4F_8 is playing the role of passivation and in fig. 4 (c) again SF_6 is etching. As a result, etching aspect ratios of 1 to 50 can be achieved. The process can easily be used to etch completely through a silicon substrate, and etch rates are 3-4 times higher than wet etching. In the following part more details are given on DRIE. Under our MEMS research group in Universiti Teknologi PETRONAS, high aspect ratio of 3 to 50 μm and complete etch through is achieved as shown in fig.5.

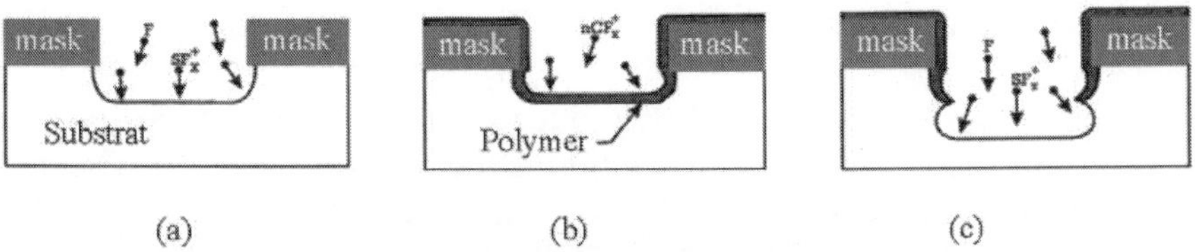

Figure 4. Illustration of etching and passivation in DRIE

Figure 5. FESEM image of a high aspect ratio plasma etching using DRIE with inhibitor

Vapor phase etching (also called Chemical dry etching) does not use liquid chemicals or etchants. This process involves a chemical reaction between etchant gases to attack the silicon surface. The chemical dry etching process is usually isotropic and exhibits high selectively. Anisotropic dry etching has the ability to etch with finer resolution and higher aspect ratio than isotropic etching. Due to the directional nature of dry etching, undercutting can be avoided. Some of the ions that are used in chemical dry etching is tetrafluoromethane (CH_4), sulfur hexafluoride (SF_6), nitrogen trifluoride (NF_3), chlorine gas (Cl_2), or fluorine (F_2) [3]. Vapor phase etching can be done with simpler equipment than what RIE requires. The two most common vapor phase etching technologies are silicon dioxide etching using hydrogen fluoride (HF) and silicon etching using xenon diflouride (XeF_2), both of which are isotropic in nature. Usually, care must be taken in the design of a vapor phase process to not have bi-products formed in the chemical reaction that condense on the surface and interfere with the etching process.

Sputter etching is essentially RIE without reactive ions. The systems used are very similar in principle to sputtering deposition systems. The big difference is that the substrate is now subjected to ion bombardment instead of the material target used in sputter deposition [4].

CMOS COMPATIBLE MICROMACHINING

CMOS compatible micromachining is now commonly known as CMOS MEMS technology. During post processing in CMOS MEMS technology, the top metal layer acts as mask to define MEMS structures. For the front side processing of the die no photolithographic steps are required. CMOS MEMS technology facilitates the wiring of MEMS structures to integrated circuits due to many metal layers. Through monolithic integration of the CMOS-MEMS devices with circuits, the parasitic capacitances is reduced and overall high performance devices is achieved. Dry post CMOS MEMS technologies are divided into two classes: Thin film micromachining technology (Thin-film Post-CMOS MEMS Technology) and DRIE bulk micromachining technology (DRIE Post-CMOS MEMS Technology).

Thin-Film Post-CMOS MEMS Technology

The process flow of thin-film CMOS-MEMS with three metal layers is shown in Fig. 6. The CMOS circuit region is designed to be covered by the top metal layer and the MEMS structures are also pre-defined by the top metal layer or the other interconnect metal layers. Fig.6 (a) shows the cross-section of the original chip after CMOS foundry fabrication, with a passivation layer on top. Two processing steps are performed only on the front side of the chip. First, the pre-defined MEMS structure is opened by etching the SiO_2 dielectric stack layer as shown in Fig.6 (b). This is done by an anisotropic SiO_2plasma etch using Tegal SS110 A DRIE system based on Bosch technology [5]. Next, an anisotropic silicon DRIE is performed using the same Tegal DRIE system as shown in Fig.6 (c), followed by isotropic silicon etch, which releases the MEMS structure by undercutting the silicon beneath as shown in Fig.6 (d). The depth of the anisotropic etch into the silicon controls the gap between the released structure and the silicon substrate. This gap should be large enough to eliminate the parasitic capacitance between the MEMS structures and the silicon substrate. In practice, it is normally on the order of 30 μm [6].

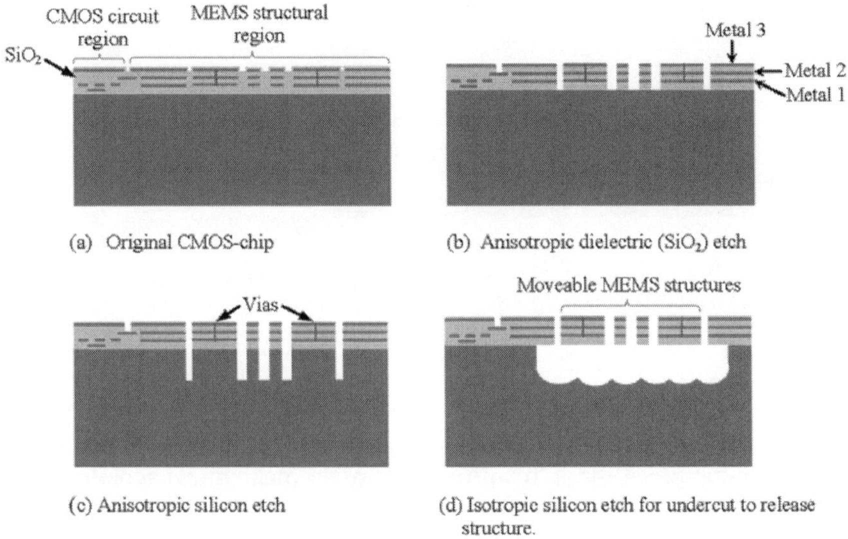

Figure 6. The process-flow for thin film DRIE post-CMOS micromachining

A resonant magnetic field sensor and piezoresistive accelerometer are examples of devices fabricated using this technology [7, 8]. This simple fabrication process yields much smaller parasitic effects as compared to the other processes such as poly MUMPs and it provides flexible wiring by using the multiple metal layers. However, there are some drawbacks in this technology, which limit the performance of the fabricated accelerometers. First, there is large vertical curling in the comb fingers of the suspended MEMS structures caused by the residual stress gradient existing in the composite SiO_2/Al layers as shown in Fig. 7. Additionally, due to different thermal expansion coefficients (TEC) of Al and SiO_2, the curled MEMS devices exhibit a strong temperature dependence, which limits their utility. Although particular compensation technology was employed to reduce the structure mismatch by using a specially designed frame, the device fabricated using this thin film process still has a stringent size limit.

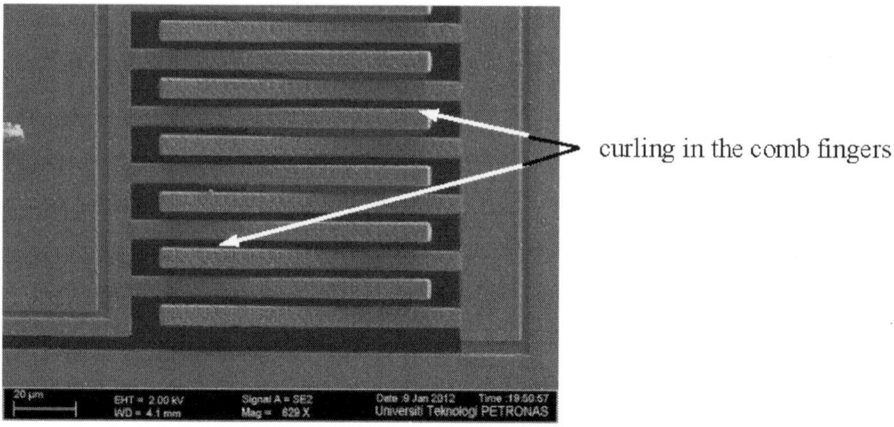

Figure 7. FESEM image of typical comb fingers fabricated using thin film CMOS-MEMS technology

Second, for sensors which require proof mass like inertial sensors fabricated using this technology, the requirement of holes on the proof mass reduces the proof mass, resulting in a lower mechanical sensitivity of the sensors. To achieve a capacitance change large enough for the readout circuit to detect, the dimension of the sensors may need to be considerably large, which is in conflict with the dimensional limit imposed by the possibility of structure curling. Third, in-plane curling of the thin film structures due to fabrication variations also limits the maximum size of the device.

Although there are some specific processes designed for low residual stress thin films [9], Farooq et al utilized the Malaysian Institute of Microelectronic Systems (MIMOS Bhd) 0.35 μm CMOS foundry processes to reduce the curling of the MEMS structures by proper design of the structures as described in [10]. Fig. 8 shows the successfully released central shuttle of magnetic field sensor fabricated using this process. It can be clearly observed that the curling in the structure has been significantly reduced as indicated by the minor curling in the fingers. In addition to the collection of technological data directly from the employed CMOS foundries, systematic post-CMOS process calibration must be conducted to characterize the process variations and their effects on the MEMS devices.

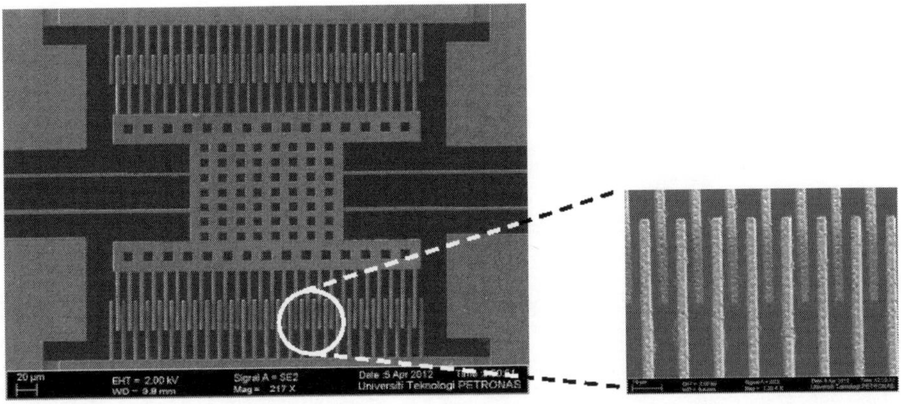

Figure 8. FESEM image showing comb fingers of the central part of a magnetic field sensor

Lastly, since the last release step is an isotropic etch, it undercuts the silicon close to the circuit and structure anchors. The ratio of the vertical and lateral etching rate is approximately 3:5. To protect the silicon underneath the circuit region from being etched away during the structure release, the CMOS circuitry must be placed far away from the microstructures, especially when a large separation between the microstructures and substrate is needed. As a result, significant chip area is wasted due to the protection margin around the MEMS structures. Since the silicon underneath the mechanical anchors of the MEMS structures is also etched away, the suspension of the mechanical structures is softened, which results in a lower mechanical performance and less robustness of the device [10].

DRIE Post-CMOS MEMS Technology

To reduce the disadvantages of the above thin-film post-CMOS MEMS process, DRIE post-CMOS MEMS technology is developed to incorporate bulk single crystal silicon (SCS) into the MEMS structures that are formed by CMOS thin films. Due to the improvement in DRIE technology, high aspect ratio CMOS MEMS structures have been exhibited as shown in Fig. 9. The achievement of maximum aspect-ratio in DRIE system is an important factor in the MEMS structure design [5]. It decides the dimensional limit of the structures fabricated using that DRIE system. Once a lateral feature to be etched is fixed, the maximum thickness is uniquely defined.

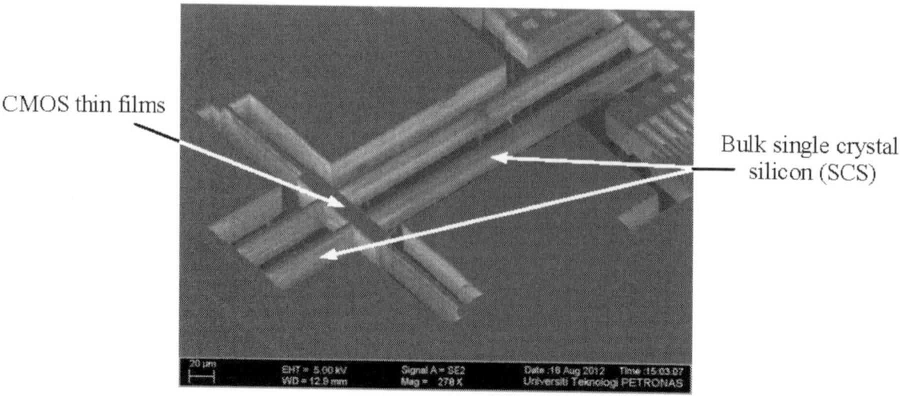

Figure 9. FESEM image showing the incorporation of bulk SCS to the CMOS thin films.

Process Flow

The DRIE post-CMOS MEMS technology branches from the thin film technology described in the previous section. DRIE CMOS-MEMS provides an approach to implement thick and flat MEMS structures with improved mechanical performance and device strength due to advantage of both flexible wiring with multiple metal layers in CMOS technology and the capability of high aspect ratio etching.Fig.10 shows the process flow of the DRIE post-CMOS MEMS technology.

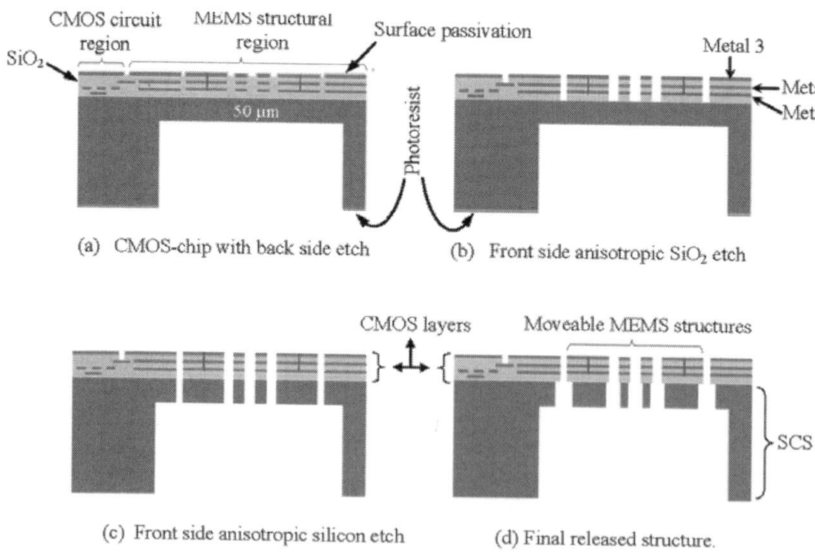

Figure 10. The process-flow for DRIE CMOS micromachining

To define the MEMS structure thickness (thin film plus substrate), the process starts with backside silicon DRIE etching by the selective application of photoresist at the back-side as shown in Fig.10 (a). The maximum thickness of this structure is limited by the smallest etching pattern on the front side of the MEMS structure and the maximum etch aspect-ratio. Next, to define the MEMS structures, an anisotropic SiO_2 etch is performed on the front side of the chip as shown in Fig. 10 (b) similar to the thin film process. In the following step, the structure is released by etching through the remaining SCS substrate. This step differs from the thin film process in that an anisotropic DRIE, instead of isotropic Si etch, is performed to release the structure as shown in Fig.10 (c). Flat and large MEMS structures can be obtained by incorporating SCS underneath the CMOS interconnect layers because the residual stresses in the SiO_2/Al thin-films are mitigated by the thick SCS, showing very little out-of-plane curling as shown in Fig. 11 where it can be seen that the sense fingers are perfectly flat and aligned.

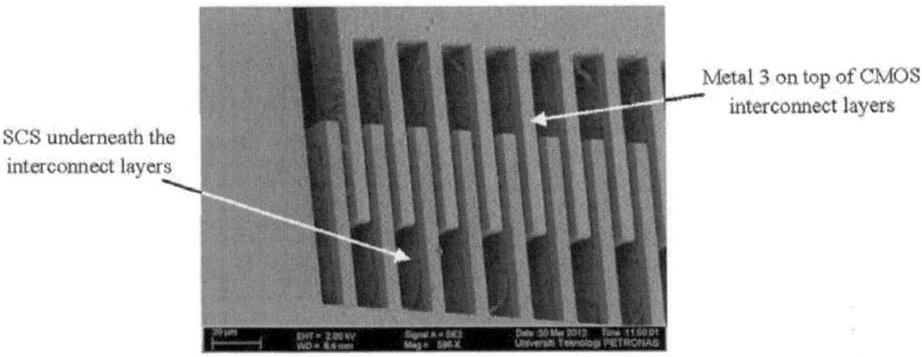

Figure 11. FESEM image showing comb fingers with thick SCS underneath

In some MEMS structures isotropic silicon etch can be added to create nice mechanical structures which only consist of CMOS thin films. A resonant magnetic field sensor fabricated under University Technology PETRONAS (UTP) MEMS Research group incorporates both CMOS thin-film layers for the beams and silicon substrate along with CMOS thin films in the side torsional coils as shown in Fig. 12.

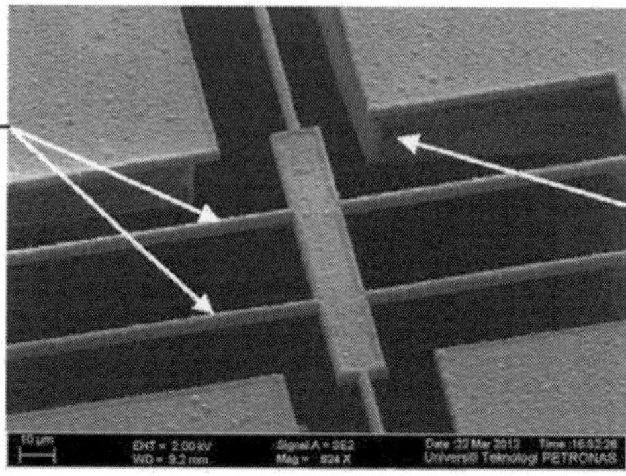

Figure 12. FESEM image of CMOS thin layers and incorporation of SCS with thin layers as MEMS structures

Double side alignment is required in the backside DRIE step to define the thickness of the MEMS structures. It is observed that normally the remaining silicon for bulk MEMS structures is on the order of tens of microns and by the substrate thinning there is no apparent circuit performance degradation. Therefore, there is no hard and fast constraints for alignment in the backside etch. A reliable sensing and actuation is achieved with the flat MEMS microstructures enabled by the DRIE CMOS MEMS technology. A resonant magnetic field sensor with piezoresistive sensing is fabricated using this technology [11]. The technology is very suitable for the fabrication of, in particular, thermally actuated micromirrors where bimorphs are used to elevate the mirror plate. Numerous resonant chemical sensors, electrothermalmicromirrors and accelerometers have been fabricated using this technology [12].

Design Rules for MEMS Devices

For CMOS circuit designs, there are different ways to break up a layout into basic elements. Designers usually divide and analyze a circuit by the function of individual components, devices and interconnect. The devices may be active (e.g. transistors, diodes) or passive devices (e.g. resistors, capacitors and inductors). The interconnects include in-plane wires, pads and vias which connect multiple metal layers. The critical information on the geometric patterns includes the size of each feature, the spacing between features and the overlap of features. These geometric data, required in the design, leads to evolutions of processes and ensures successful fabrication of microelectronic circuitry. In order to fulfill the requirements of successful MEMS device designs, processing information, which quantitatively describes the capabilities and limitations of the process under predetermined processing conditions and parameter settings, is necessary and crucial. This information is abstracted as MEMS design rules and is verified in the design environment as illustrated in Fig. 13.

Figure 13. Design rules function between MEMS designers and MEMS fabrication engineers.

This section explains how to construct concise and conclusive design rules. We used MIMOS Berhad Malaysia standard 0.35 μm CMOS design rules for the fabrication of CMOS MEMS devices and only the design rules for mechanical structures are discussed in this section. These MEMS design rules add further constraints to layouts. Similar to CMOS design rules, a complex micromechanical layout is fragmented into small pieces, and MEMS design rules contain only fundamental geometries directly related to the post CMOS micromachining process. A set of geometric constraints and design rules are then based on the final process.

Design rules are expressed in the design-rule file used in layout tools such as Cadence Virtuoso, Magicor, Mentor Graphic, layout tool etc. After all processing completion, a post-CMOS micromachined chip can be partitioned into a released area (containing free standing parts and free moving parts) and a non-released area (anchored parts) according to the mechanical behavior. The objective of post CMOS processing is the final result of releasing microstructures and maintaining critical dimensions. There are practical difficulties in monitoring each intermediate process stages. Therefore characterization of the process to extract design rules is performed at the end of the entire process flow. An overview of a typical post CMOS MEMS magnetic sensor is shown in Fig. 14.

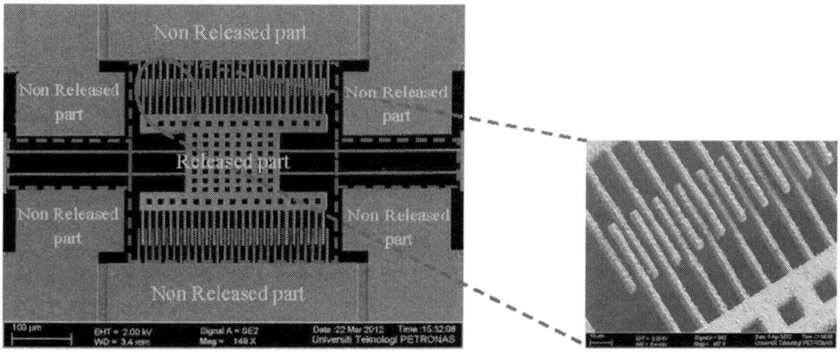

(a) (b)

Figure 14. Partitioning of a MEMS sensor for design-rule checking

The anchored part is the area outside of the red dash-line box on FESEM, whereas the released part is the area inside the red border. The close-up FESEM view of circled portion is shown in Fig. 14 (b). Sufficient paths for the gas flow are required in order to allow the reactive radicals in the plasma to undercut MEMS structures. These paths can surround mechanical structures, so that the undercut can happen from the periphery of the structures. The gas flow can also pass through structures via etch release holes. Fig. 14 illustrates various situations for reactive species undercut: from one side of the structures toward the anchor region, from both sides of beam structures, or outward through etch release holes on plates.

Theoretically, any shape of the feature which opens a gas flow path from the top of the chip to the substrate is suitable for the purpose of undercut. However, currently in CMOS foundries, Manhattan layout is preferred, or at least convenient. The majority of CMOS foundries only accept patterns with edges which are parallel, perpendicular or 45° with respect to the horizontal/vertical alignment lines. Non-Manhattan features require more and smaller rectangular patterns to approximate the geometry, and therefore, require a larger data base and longer exposure time, which leads to higher cost. Today, raster-scanning mask-making technologies such as e-beam or laser systems can overcome the difficulties mentioned above but non-Manhattan shapes approximated by a smaller raster-beam size still require more time in fabrication [13]. Following the practice of CMOS fabrication, this work only explores rectangular design rules. The major purpose of this section is to ensure the release of moving/free-standing parts and to guarantee the integration of anchor parts. Other process related rules are not included. According to situations described in Fig. 14, there have been three features encountered in the MEMS structure layout for the purpose of undercut: large openings, gaps and holes. To control the etch rates in these opening, dummy structures are used as shown in Fig.15 (a) whilefigure 15 (b) shows the fully released structure.

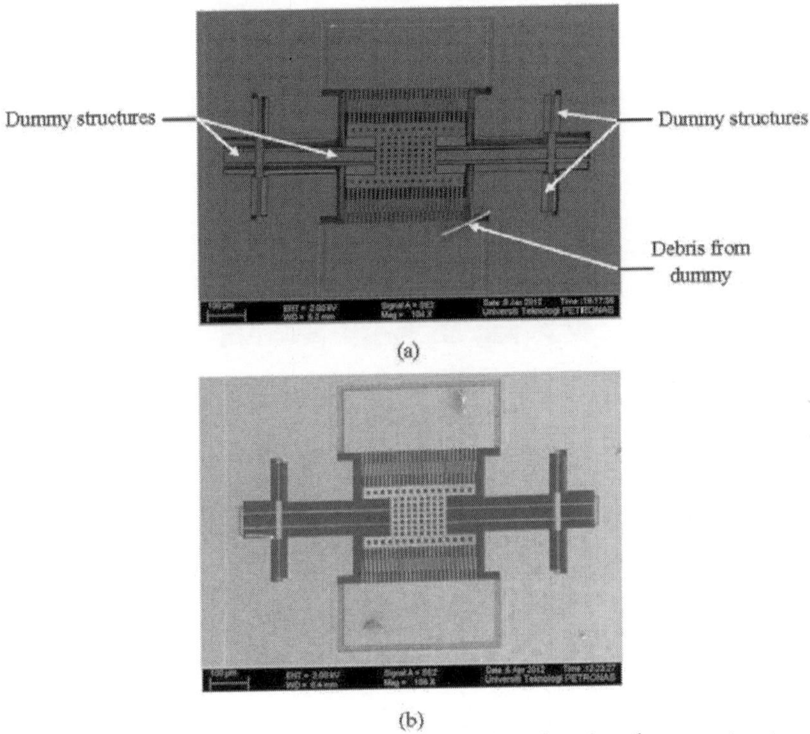

Figure 15. FESEM images of (a) unreleased device showing dummy structure at a large etch holes and (b) fully released structure.

Test structures are used to extract design rules. A large opening undercut rule specifies the maximum undercut that can be achieved by the process. Any microstructure wider than this value will not be released. This rule also specifies the minimum size requirement of the anchor layout, and the skirt for bonding pads. Bonding pads are usually on the periphery of the chip for easy access during bonding and no crossing of bonding wires on the top of the chip. Therefore, at least one side of bonding pads faces large opening and is lateral undercut during the release. The sites that experience forces during wire bonding should avoid these undercut regions and bonding pads need extra area to compensate this lateral undercut. Moreover, this large opening rules specifies the minimum space required from the edge of the etch pit to preserve Si area containing circuitry during micromachining. One way to obtain this data is to use FESEM to measure the undercut from the periphery of the chip. Another way is to use an optical microscope to check which size of square pattern has been released as illustrated in Fig. 16 (a).

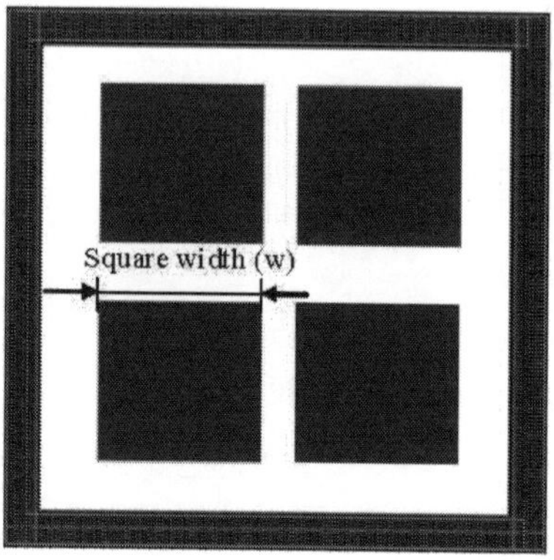

(a) The square structure illustrating the large opening undercut rule

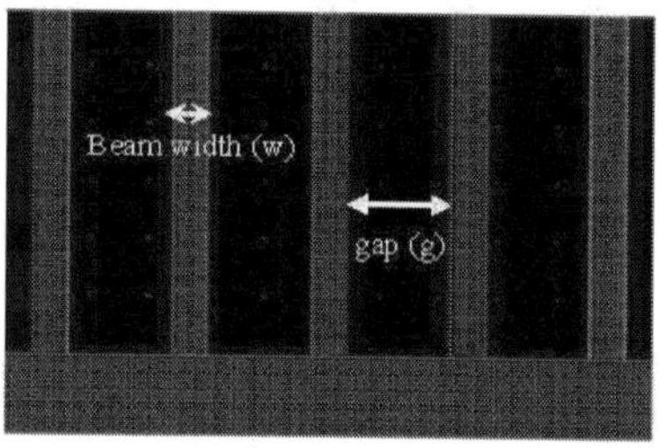

(b) The cantilever structure illustrating the gap undercut rule

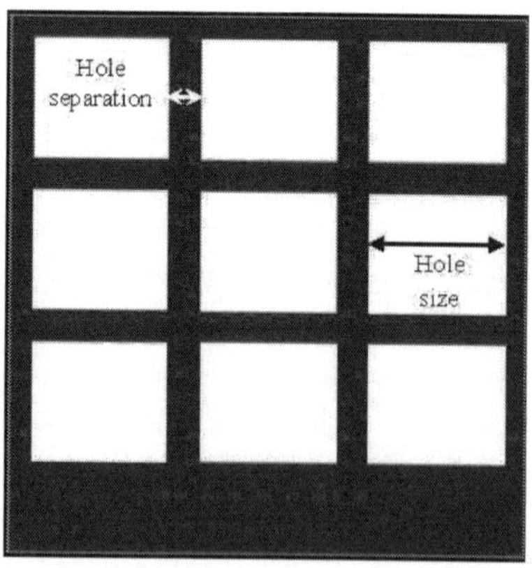

(c) The plate structure illustrating the hole undercut rule

Figure 16. Layout of three test structures used for extracting MEMS design rules for post-CMOS micromachining.

The undercut value is half the size of square width. A gap undercut rule is required for releasing MEMS structures such as a comb finger or a spring. The test structure for this rule is a set of cantilevers with the undercut mainly determined from the sides of the beams. To follow the situation in real layout, such as comb fingers, multiple cantilever beams with equal space between them are used as shown in Fig. 16 (b). The beam width (w) represents the width of the structure to be released. The gap (g) represents the separation between the structures. The beam should be long enough that the undercut from the tip of the cantilever doesn't dominate the etch. Since the residual stress induces out-of-plane curling upon release, the release can be easily detected by an interferometric measurement. A hole undercut rule is required to release plate structures. The test structure for this design rule is a set of plates with release holes. These plates are anchored by four thin beams as shown in Fig. 16 (c). Such plates should be large enough to ensure that the release of the plate is determined by the etch release holes and not from the periphery of the plate.

Similar to the cantilever test structures, the curling after the release of plates can be detected by the interferometric measurement. Extra processing time requires layouts to be modified such that the circuitry is far away from the edge of the mask, all the anchors are wide enough, and all bond pads have enough margins for bonding and probing. The spacing between the squares and the spacing from a square to the surrounding anchor mask area are the size of the square. This is to ensure that squares are uniformly etched from all directions. For each group of cantilever test structures, as in Fig. 16 (b), the widths of all cantilever beams are the same value. The spacing between beams and from tips of cantilever beams to anchor mask areas are the same value [14].

DRIE BASED CMOS-MEMS DEVICES

In this section, two devices fabricated using CMOS-MEMS bulk micromachining technology are described. The devices are a resonant MEMS magnetic field sensor and a resonant MEMS chemical sensor.

Resonant MEMS Magnetic Field Sensor

The post-CMOS process steps of the resonant magnetic field sensor using DRIE were designed to successfully release the sensor structure. The first step in the post-CMOS process was at wafer level with the selective application of photoresist at the back-side of the 8" inch silicon wafer to etch from the back-side to produce a substrate thickness of approximately 50 μm as shown in Fig.17. DRIE was performed using SS110A Tegal plasma etcher to anisotropically etch the silicon substrate to the desired thickness.

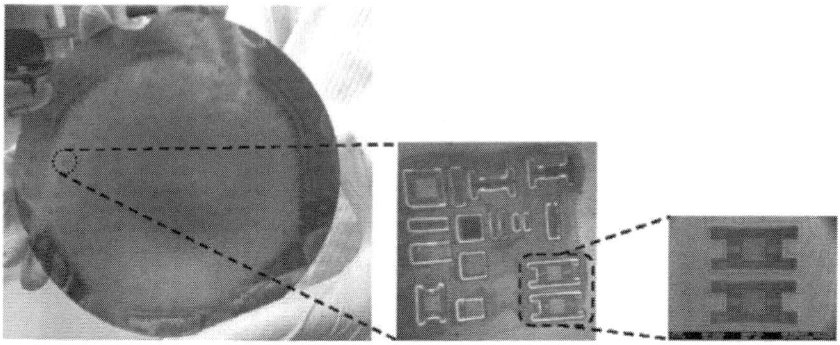

(a) (b) (c)

Figure 17. a) Optical image of back side Si etching at wafer level, (b) Backside optical image of a chip and (c) FESEM of backside of a sensor.

Optical microscope was used to estimate the thickness of the Si substrate during the etching process.Fig.17 (c) shows the back-side view of one of the sensors that has been successfully etched to approximately 284 μm depth leaving a about 50 μm thickness under the CMOS layers. From the back-side etching of the bulk silicon, the silicon etching rate was found to be ~3.66 μm/min. The second step was dicing the wafer as shown in Fig. 18 (a). Dicing of the wafer has to be done before the front side RIE of SiO_2 and DRIE of silicon is implemented. This sequence was followed in order to prevent the breakage of the released structures.

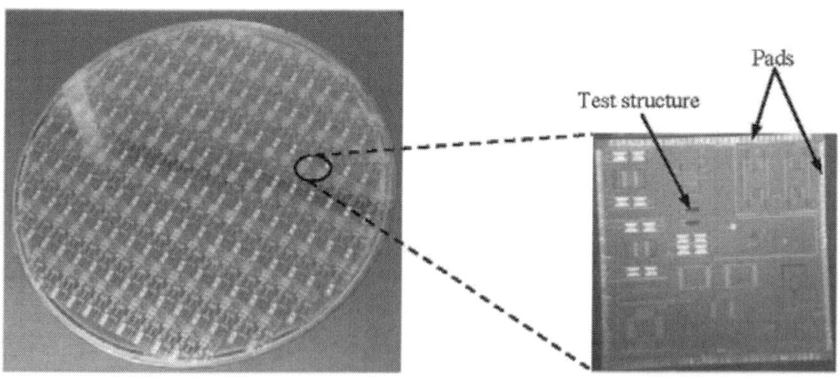

(a) (b)

Figure 18. a) Optical image of the diced wafer and (b) optical image of front side of a 5 mm × 5 mm chip.

The third process was SiO_2 RIE on the diced chip shown in Fig. 18 (b), which was performed from the front-side of the chip using Tegal plasma etcher. Front side RIE process opened the pattern of the resonant magnetic sensor by removing the SiO_2 layer and thus exposing the silicon underneath while the MEMS sensor pattern was maintained by metal 3 acting as a mask for the pattern. The process at chip level started by flipping over the chips such that the thin film was now on the front side. The chips were then attached to the 8" platinum coated carrier wafer using kapton tape as illustrated in Fig.19. After 13 minutes of etching, a gray colour (silicon colour) appeared in the trenches that indicated the completion of SiO_2 etching. The test structures to the right of the sensor beam were used to estimate SiO_2 etching rate and to avoid over-etch of the SiO_2 material as shown in Fig.18 (b). The total SiO_2etching depth was approximately 5 μm.

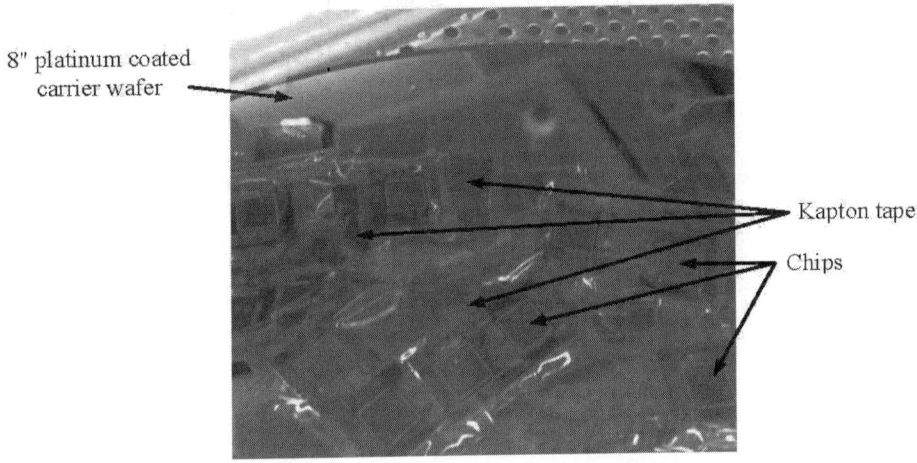

Figure 19. Sample preparation for front side etching

Next, front-side bulk silicon DRIE process step was performed using the etching rate of 3.66 μm/min obtained from previous backside DRIE etching of silicon. The thickness of the central shuttle and long beams was approximately 50 μm, which was observed under optical microscope during back-side etching. Therefore the test structures on the chip were used as reference. Fig.20 shows FESEM micrograph of the fabricated sensor with inset showing a close-up of the comb fingers and part of dummy structure close to the beams that has fallen out while others still remain intact on the substrate [10]. Additional 3 more minutes were required to fully etch-through. This happened when all the dummy structures dropped and indicated that the etched-through process was completed.

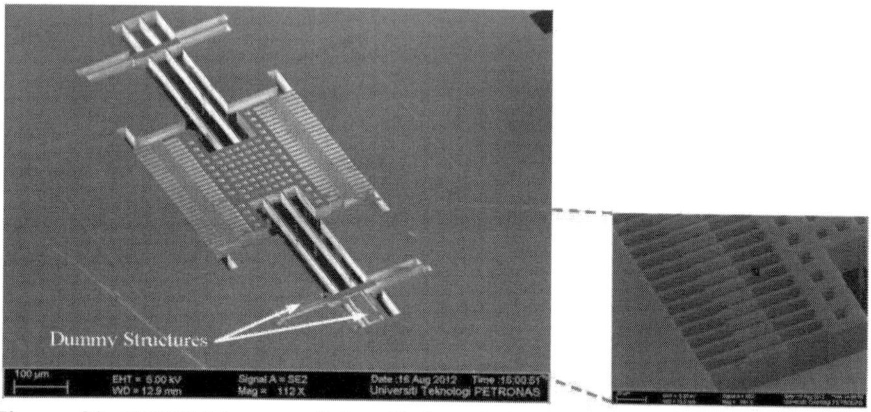

Figure 20. FESEM image of the fabricated CMOS-MEMS resonant magnetic field sensor with inset showing shuttle and stator fingers.

Resonant MEMS Chemical Sensor

In resonant MEMS chemical sensors, known as gravimetric sensors, the principle of detection of the gaseous species is based on the change in resonant frequency of the microresonator membrane or plate. This frequency change results from a change in the mass of the microresonator due to absorption/adsorption of an analyte molecule onto the surface of the active material deposited on it. Fig. 21 shows 2-D schematic diagram of the chemical microsensor device.

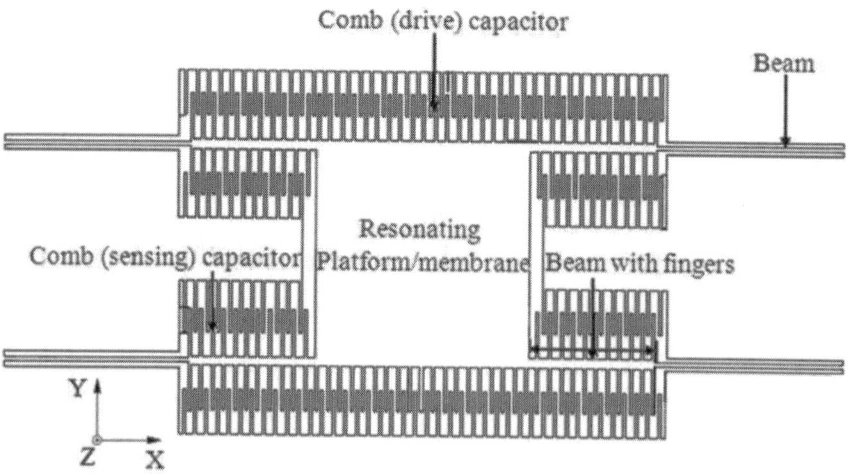

Figure 21. D Schematic of the resonant MEMS chemical sensor

The post-CMOS micromachining process steps of the resonant MEMS chemical sensor were similar to the process steps of the resonant MEMS magnetic field sensor. The process started with the selective application of photoresist at the back-side of the die around the sensor followed by sample placement on the platinum coated carrier wafer. The second step of the process was to perform back-side silicon DRIE to achieve proof mass thickness of approximately 40 μm. The same plasma silicon etcher was used to anisotropically etch the silicon substrate to the desired thickness. Fig. 22 shows the FESEM image of the successfully released MEMS chemical sensor with the inset showing a close-up view of the perfectly flat sensing comb fingers with the SCS underneath [15].

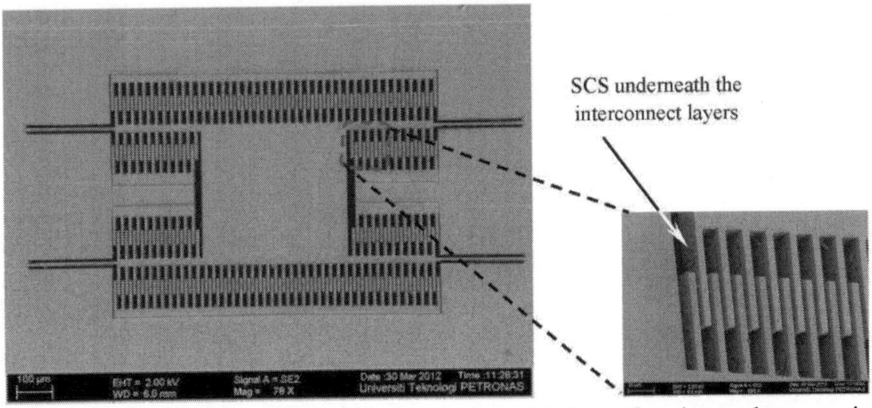

Figure 22. FESEM image of fabricated device with inset showing a close-up view of the fingers with the SCS underneath

CONCLUSION

This chapter discussed bulk micromachining technology with particular emphases on DRIE post CMOS MEMS bulk micromachining. The chapter was divided into three sections. In the first section an introduction to bulk micromachining of silicon and isotropic and anisotropic wet and dry etching was given. The second section discussed briefly DRIE post-CMOS micromachining process with particular emphasis on DRIE post-CMOS bulk micromachining process and the third and last section provides a few examples of devices fabricated by our research group using the DRIE CMOS-MEMS process. These devices were resonant MEMS magnetic field sensor and resonant MEMS chemical sensor. The aim of the chapter was to discuss and analyze practical processes involved in the design of micromechanical devices using 0.35 μm CMOS technology.

ACKNOWLEDGEMENTS

The authors would like to thank MIMOS Bhd Malaysia for facilitating the microfabrication of the sensors and MOSTI Malaysia for financially supporting this research under E-Science project No.04-02-02-SF0095.

REFERENCE

1. HongweiQu and HuikaiXieProcess Development for CMOS-MEMS Sensors with Robust Isolated Bulk Silicon Microstructures", *IEEE/ASME Journal of Micro-Electro-Mechanical Systems*, 1611521161
2. http://www.memsnet.org/about/fabrication.html
3. http://www.meengr.ccny.cuny.edu/bridge/material/joined%20files.pdf
4. http://www.mems.ac.cn/jishuwenxian/2009Etching_Processes.html
5. H Xie, L Erdmann, X Zhu, K. J Gabriel, and G. K Fedder, Post-CMOSprocessing for high-aspect-ratio integrated silicon microstructures," Journal of Microelectromechanical Systems, 11200293101
6. G. K Fedder, S Santhanam, M. L Reed, S. C Eagle, D. F Guillou, M. S. C Lu, and L. R Carley, Laminated high-aspect-ratio microstructures in a conventional CMOS process," The Ninth Annual International Workshop on Micro Electro Mechanical Systems, San Diego, CA, 19961318
7. F Ahmad, J. O Dennis, M. H. M Khir, and N. H Hamid, A CMOS MEMS Resonant Magnetic field Sensor with differential Electrostatic actuation and Capacitive sensing," Advanced Materials Research Journal, vols 403-408, 420542092012
8. M Haris, and H Qu, A Cmos-mems, Piezoresistive Accelerometer with Large Proof Mass", IEEE NEMS 2010Xiamen, China, Jan 2010.
9. H Xie, Gyroscope and micromirror design using vertical-axis CMOS-MEMSsensing and actuation," Ph. D. Thesis, Department of Electrical and Computer Engineering, Carnegie Mellon University, Pittsburg, PA, 2002
10. F Ahmad, J. O Dennis, M. H. M Khir, and N. H Hamid, Lorentz force based CMOS Resonant Piezoresistive Magnetic Sensor with differential electrostatic actuation" MY. Patent, Jan.16, 2012
11. FarooqAhmadDesign and simulation of mechanical behavior of MEMS-based resonant magnetic field sensor with piezoresistive output", 2010International Conference on Mechanical and Electrical Technology, 09/2010
12. G. K Fedder, Tri-axial high-g CMOS-MEMS capacitive accelerometer array", 2008IEEE 21st International Conference on Micro Electro Mechanical Systems, 01/2008.

13. X Zhu, Post-CMOS Micromachining of Surface and Bulk Structures," Ph. D. Thesis, Department of Electrical and Computer Engineering, Carnegie Mellon University, Pittsburg, PA, 2002
14. Brand, CMOS-Based Resonant Sensors", IEEE Sensors 2005
15. Y Ahmed, J. O Dennis, M. H. M Khir, and M. N. M Saad, A Method, Apparatus and Sensor for Reversible Detection of Analyte Molecule" MY. Patent, Dec.16, 2011

CITATION

John Ojur Dennis, Farooq Ahmad and M. HarisKhir (2013). CMOS Compatible Bulk Micromachining, Advances in Micro/Nano Electromechanical Systems and Fabrication Technologies, Assistant Professor Kenichi Takahata (Ed.), ISBN: 978-953-51-1085-9, InTech, DOI: 10.5772/55526.

Index

I

L

M

N

P

R

S

T

U

V

W